||||||||||||||||||||||||||||

# VEHICLE DESIGN

*Vehicle Design* guides readers through the methods and processes designers use to create and develop some of the most stunning vehicles on the road. Written by Jordan Meadows, a designer who worked on the 2015 Ford Mustang, the book contains interviews with design directors at firms including Fiat Chrysler Automobiles, Hyundai Motor Group, and Ford Motor Company, among other professionals. Case studies from Ford, Mazda, and Jeep illustrate the production process from research to execution with more than 245 color behind-the-scenes images in order to help readers create vehicles drivers will cherish.

# VEHICLE DESIGN

## AESTHETIC PRINCIPLES IN TRANSPORTATION DESIGN

**JORDAN MEADOWS**

Routledge
Taylor & Francis Group

NEW YORK AND LONDON

First published 2018
by Routledge
711 Third Avenue, New York, NY 10017

and by Routledge
2 Park Square, Milton Park, Abingdon, Oxon OX14 4RN

Routledge is an imprint of the Taylor & Francis Group, an
informa business

Library of Congress Cataloguing in Publication Data
A catalog record for this book has been requested.

ISBN: 978-1-138-68559-8 (hbk)
ISBN: 978-1-138-68560-4 (pbk)
ISBN: 978-1-315-54314-7 (ebk)

Designed and typeset by Alex Lazarou (alexlazarou@aol.com)

# CONTENTS

# FIGURES

j.mea

FIGURES

# ACKNOWLEDGMENTS

It would be impossible to acknowledge all who made this book possible. However, special thanks must be given to some key supporters who were actively involved in its creation. Each of the interviewees are friends, colleagues past and present, and truly inspirational for me as a designer. I thank Ralph Gilles, Angela Weltman, Robert Davis, Freeman Thomas, Peter Schreyer, Raj Nair, and Moray Callum all for lending time and personal wisdom in support of this book.

Special thanks also to the Design, and Public Relations teams of FCA, Kia/Hyundai, Mazda, Ford Motor Company, and Tesla. Without them and the special help from Dianna Gutierrez, Vera Uhle, Jeremy Barnes, Francesca Montini, and Craig von Essen, much of this would not be possible. Special thanks also to the ArtCenter College of Design, and to Christine Hanson, Stewart Reed, Jay Sanders, Ken Nagasaka and the entire Transportation Design Department.

I also must personally thank John Clinard and Tracey Grant, two friends with an amazing amount of knowledge about the world of cars and publishing. Their guidance, mentorship, and technical/editorial assistance were truly invaluable!

Very few vehicle designers publish so I must also name those who inspired me to give it a try. Jerry Hirshberg, Chris Bangle, Stuart Macey, Geoff Wardle and Daniel Simon, along with a special personal acknowledgment to J. Mays, all showed me the value of sharing an intellectual point of view beyond one's contributions to a manufacturer.

Finally, I must also cite my family, friends, and colleagues who challenged me to always learn, grow, and create. Their love, friendship and support have enabled me to have a very special career and share some of what I've learned from it in the pages of this book.

# STATEMENT OF AIMS

*Vehicle Design: Aesthetic Principles in Transportation Design* pulls back the curtain on the secretive world of vehicle designers to reveal the methodology and processes used in the conceptualization and development of some of the industry's most visually stunning vehicles.

Turn the first page and enjoy a rare deconstruction of how professionals satisfy the most complex and demanding technical requirements while delivering an emotionally enriching experience. It provides an in-depth view of what was previously considered a dark art among mainstream designers, something akin to sheet-metal sorcery. There aren't many in-depth books about this relatively small, focused, and dedicated profession. There are even fewer written by a designer intended for all designers … *Vehicle Design* arrives at a critical time as the transportation industry is now on the cusp of a dynamic new era in mobility provided by zero emissions technology and autonomy.

Designer Jordan Meadows has packed the pages with first-hand insight and case studies from a career spanning two decades of creative involvement at some of the world's largest and most influential auto manufacturers. He also provides exclusive in-depth interviews from leading experts and Design Directors of significant automotive brands renowned for top-level design. They reveal professional experiences, secrets of success, failures, and moments of magic along the journey of creating unforgettable cars.

However, this book isn't just for automotive designers, engineers, and marketers; it has relevance and appeal to any creative-minded student or professional who is committed to the creation of meaningful images, objects, spaces, and experiences. It is also "must read" material for disciples of car culture and automotive enthusiasts of all ages and backgrounds. The images, words, and ideas are sure to empower and leave you ready to "start your creative engines."

# INTRODUCTION

In life, most of us have a job, a career, and if we're lucky, sometimes a calling. Jobs can help sustain you, and often are a great way to learn and maintain a skill or trade. If you're unlucky, sometimes they can be a total bore. Careers can last long periods of our adult lives, and whether we enjoy them or not can be an integral part of our day-to-day lives. Whether you work to live, or live to work, careers are always part of the equation. And then some of us, a very rare few of us, are called toward a particular task. Some deep drive compels us to perform a function. For this group, in a sense there isn't a choice in the matter. It's about doing it or being relegated to seriously antisocial and unfortunate behavior!

For as long as I can remember, I have had a deep, compelling drive for making things. Many children have an aptitude in one direction or the other. Some are called to dance. Others are athletic. Some feel a natural inclination to help others. We can probably blame or thank our parents for the nature and nurture aspect of our upbringing. For me, it was the unrelenting drive to create things. Somewhere along the line, shortly after walking, I discovered the magic of vehicle mobility. From a personal growth and development standpoint this was probably related to freedom and independence. Crawl, walk, roll, ride, etc. ... Whether it was the first bicycle, skateboard or anything with wheels to that end, the fascination with transportation was a calling that I experienced at a very early age. So for me, the link between creating things and vehicles, no matter how rudimentary, was quite natural.

Many professional vehicle designers and students of the craft share a very similar discovery of this unrelenting drive to create things that move. And for us it is a very deep calling.

This book was created by a designer for designers who understand that calling. It's meant to serve as a guide through the creative process of generating vehicle concepts. As with most creative endeavors, the process is in fact a journey. We begin with an intended destination or an ideal of where we want to be, resulting in a product or goal. However, going through that process can take us to places that we could never imagine along the way. And ultimately, the true value is not the destination or even the finished product but rather the experience of learning along the way that is both valuable and transformative. In the creative field, no two projects are the same. The thrill and excitement of discovery are undeniable. Perhaps this is the reason why it's often easy to remain passionate about making things throughout one's life.

Given that no two projects are the same and that each individual designer will always bring their own personal point of view to a project, this book is not meant to serve as a universal template for people undertaking this task. It's not a vehicle design bible or manifesto. It is also not about drawing techniques, digital tools, or practices that an individual can develop and apply to their own effect. Rather this book is meant to offer proven techniques for interpreting the twists, turns, challenges, and opportunities that may come up along your quest to develop vehicles with emotional and aesthetic appeal.

The information in the book is presented in three main modules. The first is about identifying interest and targets. For students, this can be honestly coming to terms with strengths and weaknesses and areas for investigation in their own personal portfolio. For working professionals, this can be business opportunities and defining a direction for their company or organization to conduct a study. For passionate individuals, this can be the process of developing a framework to help structure their invention or passion project. Whether the goal is for academic and learning purposes, or for a great business opportunity, or just a project that you've always thought was cool and wanted to do, the first module is all about defining and articulating *WHY* it needs to happen.

The second module is about identifying external factors. There is an old saying that necessity is the mother of invention. Good design and invention go hand-in-hand. Whether you're a novice student of design or a seasoned veteran of the creative game, it's always essential to understand present and future market conditions. It is also crucial to be able to identify short-term trends and macro movements that may indicate an unmet customer need. And satisfying that unmet customer need from a pragmatic, emotional, and visual perspective is what good designers always seek to do. In basic terms, this module is all about deeply understanding *WHO* the design is for.

In the real-world context these two modules can in some ways be intertwined and symbiotic. In a certain way, it's the chicken and egg story. What comes first: the customer's needs or the product solution that addresses them? One can make the argument in either direction. History has shown us that true visionaries often provide solutions that customers are not aware that they "need" at the moment BUT will become crucial later as their context and usage scenarios evolve. For this reason, it can be risky to base a design solely on existing consumer assumptions. Conversely, not having any regard for customer wants or needs can yield a product that is irrelevant or obscure.

The third phase is dedicated to giving shape and physicality to a future vehicle that satisfies unmet needs, and delivers a unique and highly emotional experience for the user. This is a blending of key insights developed in modules 1 and 2, and is truly about the execution and delivery based on your personal design philosophy and values. Module 3 is all about developing a tangible product to give the user *WHAT* they want or need. Understanding these three elements is the foundation for any design project: first, business goals and targets; second, marketplace opportunities and external factors; and, finally, appropriate execution and delivery of an experiential or physical product solution.

In short, successful design is the result of a very clear understanding of *WHO* the user is, *WHAT* it will take to fulfill their needs, and *WHY* your offering is different and more compelling than the competition's.

The learning and insights shared in this book are based on over two decades of personal experience in the automotive industry. It also contains informed opinions from influential figures in the broader world of cars and transportation design. Together we offer case studies and stories of vehicles that were made in the past. However, the focus and intent of this book is looking forward. The aim is to provide insight into how the vehicle design industry gives emotional meaning to products so that you, the reader, can build on these techniques and strategies to create fascinating and memorable design statements for the future.

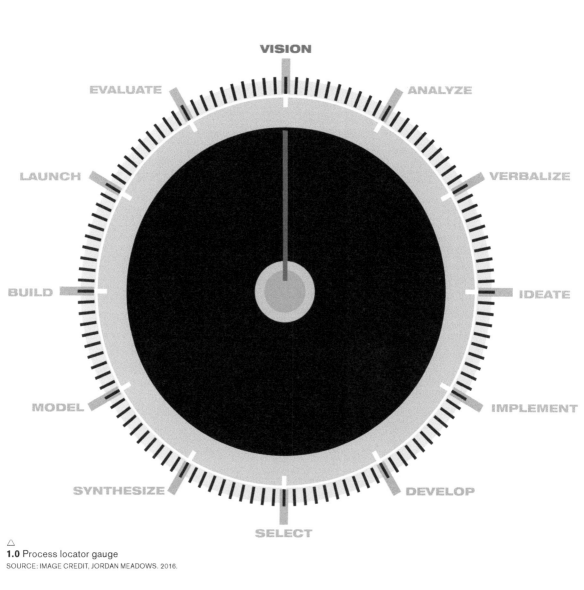

△

**1.0** Process locator gauge
SOURCE: IMAGE CREDIT, JORDAN MEADOWS. 2016.

**VISION**
LET THE JOURNEY BEGIN!

**CHAPTER 1**

# VISION
## Let the Journey Begin!

## Identifying Opportunity, Defining a Vision, Setting Targets

The design process is very much a journey, and engaging in it can be seen as embarking on an exploration. In this regard the old proverb that says the longest journey begins with the first steps holds a good deal of truth. This chapter is about taking those first steps.

Whether the reason for the exercise is user-inspired, business-oriented, or just plain born of personal passion, there is always a vision! (See Figure 1.0.) Knowing how to achieve it can be quite cloudy or undefined at times. Other times it can be crystal clear, succinct and laser-focused. But in any case, one of the first tasks of the designer is to articulate a vision and identify how it relates to an opportunity to give a user a meaningful experience.

It's helpful to engage in these first steps with a completely open mind. In the beginning of the project almost anything goes. Ideas that may seem radically divergent can lead to unexpected solutions. In a sense, this stage of the process is the most thrilling because so much can happen. Beginnings are inherently exciting because it's the point in the journey when there is plenty of room to dream and the possibilities are infinite.

Even if you've been given very strict constraints, a designer has a duty to propose solutions that work within the constraints as well as offer ideas that challenge them. In professional scenarios there will be established parameters that define a brief. This will include a set of assumptions to work within. Regardless of how pre-described or open-ended that brief may be, it's often beneficial to initiate the earliest stages of a program with a good understanding of that brief while also identifying a margin and scope of opportunity just beyond it. At the beginning of any project always take the opportunity to brainstorm. Use it as a semi-formalized exercise to allow ideas that fall within the brief and just beyond it to be discussed and entertained. Whether it's conducted with a group, or as an individual, it's a great way to kick off a project and allow the creative flow to begin. In a sense it's very much about dreaming: Diligent, semi-focused day dreaming about a possible solution at this stage can literally and figuratively be your job!

## Verbalizing the Vision: Is a Mental Picture Worth 1000 Words?

There are many techniques for brainstorming and different schools of thought on how to make the exercise truly effective. In any case, almost any project will need to go through a phase of examining the widest range of possibilities, defining the creative vision, and setting targets (see Figure 1.1). As a design takes shape, the amount of constraints will rise, so in the early stages it's fun to do some lucid dreaming about possibilities of what might be achievable. Most of the established techniques for brainstorming involve an individual or group figuratively draining their minds of possibilities, and jotting these ideas down on Post-it notes. It's helpful when capturing these ideas to be very succinct. Sometimes a great idea can be described like a joke with an amusing punch line. In many cases, great thought starters or issues that are posed in the form of a question can kick off idea generation and foster creative discussion.

Whether you're working with a group or as an individual, there are some key things to keep in mind: Establish a time limit and clearly identify measurable and specific goals for the exercise in terms of how it relates to the brief. Allow your mind to run wild with possibilities and refrain from judging any of them until after the exercise is complete. Even nonsensical or seemingly absurd ideas can trigger a fruitful discussion. And, finally, capture and organize thoughts in a way that's easily shared. This will allow you to organize the thoughts in terms of relevance, and also identify how different solutions might relate to each other and offer some synergies and interesting combinations.

In a sense, you don't need thousands of words; you only need a few words to verbalize the vision! Dreams and ideas become thoughts. Thoughts transfer into words. Words eventually become a series of actions. Actions involve decisions, and decisions eventually lead to design!

Every project begins with a series of thoughts and visions or feelings. Every designer has a responsibility to dream about an ideal solution if only for a short amount of time in the beginning of a program, regardless of the constraints involved. These are the first steps in embarking on any creative journey, and giving relevance and meaning to a design!

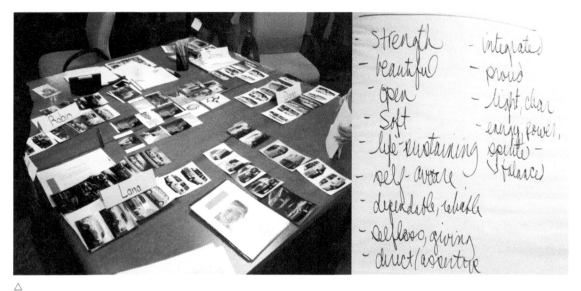

△
**1.1** A typical brainstorming session involving imaginary personas, which cars they might enjoy, and keywords that express the emotional and pragmatic targets for the intended user
SOURCE: IMAGE CREDIT, JORDAN MEADOWS.

## Identify and Analyze Strengths and Weaknesses and Opportunities in Your Portfolio

We all have strengths and weaknesses regarding the work that we do. In the initial phases of our careers as designers, it's best to gain as much experience as possible and present ourselves as well-rounded and able to take on a multitude of different challenges. One of the first steps in identifying an opportunity or a potential new project for your portfolio is to analyze what you already have, and examine where the deficiencies might be. For example, we all know designers who are in love with sports cars; and, as you would think and expect, they only have sports cars in their portfolio. And while their enthusiasm, focus, and dedication are enviable, the fact of the matter is the scope of their portfolios and their range of focus are limited. In fact, some successful sports car manufacturers might agree.

Even the most legendary sports car manufacturers now are also producing vehicles such as SUVs and sedans. So it stands to reason that it would be beneficial to investigate other opportunities for different types of vehicles. Having a well-rounded portfolio that shows that you can tackle different types of vehicles will always be beneficial in the job hunt.

For individuals who are currently in a professional setting, variety is also good. Sometimes as creators we can get stuck, pigeonholed, or overly focused. In this regard, designers are very similar to actors, and becoming type-cast in a particular role can be a real danger. Working on different types of products periodically keeps us fresh from a creative standpoint. Because the lead-time and development for vehicles are quite long, professional designers can go many years focused on one type of product, and in some cases one aspect of that product ... It's not difficult to get burned out in this regard. So whenever possible, creative individuals are encouraged to stretch their wings and take on as much variety as they can.

To a certain extent this is quite true with design-based organizations. Large organizations and design groups are reflective of their creative employees and vice versa. And in this case, having a diverse portfolio can be beneficial from a business perspective. Offering different types of products in an optimal range can allow a group to have a broader presence in the marketplace. Almost any business manager will tell you diversity is a good thing. This holds true for stocks and investments, and can be said for hiring practices too. It also rings true for the types of design assignments that you take on. Balance and some degree of diversity are a key to success!

A common methodology for achieving the right balance and analyzing what needs to happen in your portfolio involves identifying strengths, weaknesses, opportunities and threats: this methodology is known as a **SWOT** analysis (see Figure 1.2).

This is a popular method for looking at the things that one does well and identifying them as strengths. Conversely, weaknesses or areas that need improvement are also outlined. While looking inward and identifying these dynamics, one can also take an external view and define opportunities in the marketplace that are unaddressed. Finally, this methodology also allows for some understanding of competitive actions, threats

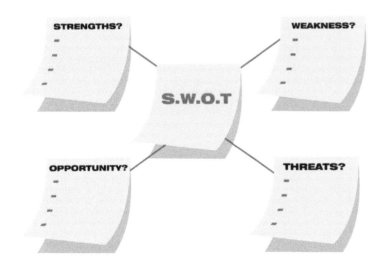

△
**1.2** SWOT analysis can be a fluid exercise undertaken with Post-it notes, and adds value to any brainstorming session
SOURCE: IMAGE CREDIT, JORDAN MEADOWS.

△
**1.3** The Dodge Kahuna concept, front view, featuring a dominant truck-like Dodge graphic, robust crossover stance, and a pillars-less window aperture in which all glass drops for open air cruising
SOURCE: IMAGE CREDIT, FCA DESIGN.

from others, and external forces that may have a negative effect on personal goals and business objectives.

While SWOT is a great strategy for self-analysis, it does have its limitations. For example, what may seem like a strength could potentially be a liability. The marketplace and competitive landscapes in business and design are fluid, volatile, and ever changing. What can be a strength at one point can quickly be turned into weakness. This holds true for organizations in a broader sense, but also for individuals.

From an individual perspective, when designers position themselves as stylists, quite often they can become very popular. If their work gains traction, it can be viewed as in vogue. When a designer is viewed as an arbiter of style, their understanding of short-term trends can potentially be seen as a strength. Having said that, if they become overly associated with a unique style or trend, they become irrelevant as soon as that look or style moves on or changes. They've effectively turned their strength into a weakness.

For organizations that are overly focused, one-dimensional, or hedge their success on one aspect, the same is also true. Understanding that a SWOT analysis has its shortcomings leads us back to the idea that diversity is very important. Organizations will often experiment with a **concept car** that addresses their strengths and weaknesses in the interest of diversifying their portfolios (see Figures 1.3 and 1.4). The Dodge Kahuna was such an exercise I was responsible for, while a designer at Chrysler. It was the result of a brainstorming

**1.4** The Dodge Kahuna concept, rear view. Featuring a technical progressive lamp graphic that accentuates a forward-swept rear profile hearkening back to classic American people movers. Complemented with a fully retractable roof and bespoke accessories for surfboards.
SOURCE: IMAGE CREDIT, FCA DESIGN.

session to examine the shifting people-mover segment that the corporation dominated with its minivan offerings when the market was beginning to shift toward cross-over vehicles. Individuals can use a similar strategy with their own personal endeavors to remain competitive. With this in mind, one can brainstorm possible project directions and begin to chart a path based on one's personal vision as outlined earlier in this chapter. But it's very important to do this with a clear understanding of one's current context, and a feel for how key aspects of it may shift or change.

## Research Examples of Personal Design Manifesto and Design Movements

The drive to create new statements as a designer is very strong. Accomplishing something that resonates with the user is appropriate, and making it fit for task is even more of a challenge. None of this can be accomplished without a fair understanding of what already exists, and why it was successful or failed miserably before. In this regard the various age-old sayings about repeating history ring true; those who don't have an understanding of why or how

something was designed before, are doomed to repeat it.

For this reason, an essential step in the journey is to have a good understanding of the history of the type of object that you want to design. You cannot move forward without looking back. When searching for an inspiration for an alternative to the normal minivan, the team at Chrysler investigated classic American wooded wagons as a source of inspiration; this thinking led to the Dodge Kahuna concept (see Figure 1.5). Sometimes you have to take a few steps back to get a running

△
**1.5** The Dodge Kahuna concept, debuted at the North American International Auto Show, is an example of a design study that intentionally sought to elicit strong emotional reactions from its viewers. The creative team generated the concept through strategic analysis of strengths, weaknesses, opportunities, and threats of the brand that was known to be the design leader in the people mover segment
SOURCE: IMAGE CREDIT, FCA DESIGN.

start forward. And because design is an inherently competitive business endeavor, it's crucial to also look side to side to have an understanding of what your peers and competitors are doing.

Corporations and groups have various ways of doing this via **competitive benchmarking** and contemporary trend analysis. Individual designers can partake in similar exercises on a smaller scale. However, in both regards, in the initial phase of beginning a project, taking a broad overview to position oneself in the competitive landscape is essential. This is crucial to gain a 360° situational awareness of the space that you're entering, knowing what took place in terms of history, understanding the competitive landscape from a contemporary perspective, and then establishing goals and attributes to work toward moving forward.

With this in mind, it's helpful to identify who your heroes are and why. No creative person exists in a vacuum without influences and sources of inspiration. It's also true that designers tend to put out what they take in. Give yourself a certain type of inspiration or input: it will result in a corresponding output. For example, it's probably safe to say that Jonathan Ive, Apple computers' influential leader of design, was inspired by the work of legendary industrial designer Dieter Rams. So many of his 10 principles of good design are evident in all of Apple's products:

> **Good design is innovative. Good design makes a product useful. Good design is a static. Good design makes a product understandable. Good design**

**is unobtrusive. Good design is honest. Good design is long-lasting.**

> **(Dieter Rams, 10 principles of good design)**[1]

In fact, so much of the visual DNA in Apple products has its roots in the clean Teutonic efficiency pioneered by Rams' work for Braun. To dig even deeper, Rams' work was rooted in, and inspired by, the Bauhaus and Modernist movements. These manifestos, developed and popularized by Walter Gropius, Mies van der Rohe, and others, underpin a good deal of architecture and product design in the Modernist and minimalist movements of the mid- to late-twentieth century.

When setting out on the first steps of a project it's helpful to identify design manifestos and movements that have occurred as a source of visual inspiration for the creative journey you want to take. Understanding these movements can also be valuable if you'd like to rebel against them and create your own. Whether it's "modernist," "deconstructivist," "surrealist" or "abstract impressionistic," most of these movements were outlined in a manifesto or verbal proclamation of the ideology. And while they can be described in verbal terms, and their socio-cultural context is important, the immediate use for designers can be the visual inspiration and imagery they yield. Case in point, a vehicle derived from the thinking of mid-century Modernism will look very different from a vehicle that subscribes to a deconstructivist manifesto. The Dodge Kahuna may be described as fitting a post-modern philosophy, or possibly retro-futuristic (see Figure 1.5). This was a strategic move to court

controversy and rebel against the contemporary styling norms of its mainstream minivan competitors.

Identifying a reservoir of visual inspiration is essential in the first steps of the project. Utilizing and applying that visual imagery effectively is a subject covered later on in this book. Just as it's important to source inspiration that relates to your vision and ideals, it's also crucial to be mindful of imagery to avoid. And just as every journey has heroes, teachers, and mentors, there are also villains and demons to be avoided.

An extremely valuable exercise at this point is establishing and verbalizing what you don't want to do, and identifying results you will try your hardest to avoid. Most designers, whether they are die-hard enthusiasts or not, have a vehicle that they can appreciate and enjoy looking at. Conversely, almost everyone can identify a vehicle that they really don't care for and find visually awkward, difficult to look at, and unfortunate from a design perspective. But before we let the bashing begin, it's worthwhile thinking about what it is that makes those vehicles unsuccessful. The reasons that contribute to a product being viewed as trash or treasure are important to consider and analyze. Quite often it's the result of misguided practices and assumptions in the product creation. Or sometimes the issue lies outside of the designer's scope of influence. In either case, these are the real-world challenges that anyone creating objects will face, so it's crucial to anticipate the demons and obstacles lurking along the journey.

**Identify a Customer and Market Opportunity Based on an Emotional Experience**

In this chapter we've discussed various methodologies for verbalizing the vision and getting started in the process of designing a unique and meaningful vehicle. They include open brainstorming and verbalizing a designer's personal vision. They also include identifying and analyzing strengths and opportunities in your portfolio. There is always a need to research historical and contemporary examples of what, and what not, to do. It's also helpful to source visual inspiration and name influential figures and movements. However, the most

△
**1.6** Design, being different from art, is a blend of the personal vision that seeks to offer a unique experience for a user, the simple equation can form the foundation for any design exercise
SOURCE: IMAGE CREDIT, JORDAN MEADOWS.

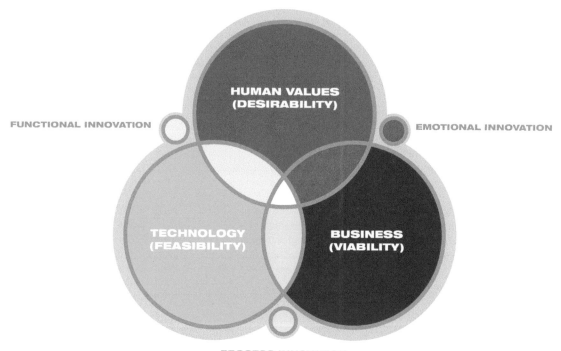

△
**1.7** Industrial design powerhouse IDEO developed this figure that depicts an ideal blend of three key attributes to deliver resonant and meaningful design work[2]
SOURCE: IMAGE CREDIT, IDEO.

crucial step at the beginning of any design process is to identify and understand a potential user, and define a set of ideal emotions you would like them to experience (see Figure 1.6).

In business and design publications, much has been written about the importance of making an emotional connection with a user. For designers and creative types this is a bit of a no-brainer: in fact, a basic instinct is to create objects that people feel a connection with. After all, what's the sense in doing something that nobody finds really cool? It may seem obvious that successful products and services are ones that are designed to appeal on an emotional level; nevertheless, it is easier said than done. It requires consistency, great timing, taste, and attention to detail. True mastery of this is something that all working professionals should constantly aspire toward throughout the course of their careers.

For the sake of the exercise we can define a meaningful experience as one that is memorable. We can also say that an experience is the act of encountering or having contact with facts, objects, actions, or events. Deconstructing and analyzing any meaningful human experience can yield key insights on which direction to take a project from asking a few key questions.

Honest and true design always comes from the heart. We know our own hearts and emotions better than anyone! For the sake of this exercise, recall a meaningful and memorable experience you personally had with a vehicle:

- Which aspects of it were pleasurable?
- Which aspects were challenging, awkward or difficult?
- In a best case scenario, how would that experience unfold and what emotional response would you have as a result?

Describe in a qualitative manner how another individual would feel partaking in this experience in an ideal way. Through constantly referring to this series of questions we can establish aspiration and desired outcomes. By transferring these questions to a potential user's interaction with our work, we launch our design projects with a sense of empathy, and empathy is the most powerful and transformative component of any creative process! There are, however, at least two other components to consider in our work as designers. Business viability and technical feasibility are also crucial considerations, and reconciling them will be part of your creative journey moving forward (see Figure 1.7).

IIIIIIIIIIIIIIIIIIIIIIIIIIIIIIIIIIIIII

# Q&A

**RALPH GILLES**
EXECUTIVE VICE PRESIDENT OF
DESIGN, FCA

*Ralph V. Gilles was appointed
Head of Design, Fiat Chrysler
Automobiles Global, in April 2015.
Ralph has also served as the
Senior Vice President, Product
Design and President and CEO,
Motorsports, FCA–North America;
President and CEO, SRT brand;
and President and CEO, Dodge
brand for FCA US LLC. He was
named Vice President, Design in
September 2008 (see Figure 1.8).*

*Ralph has graciously agreed to
share some key insights on how
his organization initiates projects,
identifies opportunities, and gets
the ball rolling!*

△
**1.8** Ralph Gilles, Head of Design, FCA Global
SOURCE: IMAGE CREDIT, FCA DESIGN.

**Question 1: Who are your
personal design heroes/who do
you most admire, and why?**

▶ When I was young, just
learning about design, I really
didn't know who the movers and
shakers were. It wasn't until much
later that I got to understand who
people like Harley Earl were—
people who got design to have
a prominent seat at the table in
a corporation. There are other
examples of that. A lot of people
like to use Apple—but in most
successful companies, the Design
department is aligned to the top of
the house, so to me, people who
do that are very important.

Later I learned one of my heroes
was Guigaro. When you think of
Italian design, you think of these
design houses: Guigaro and
Italdesign, Pininfarina design,
Bertone, etc., you had all these
great independent studios. With
Guigaro and people like him, I
was amazed at how prolific they
were, that they could design an
Alpha Romeo one day, and a

Volkswagen the next, and even taking on industrial design and product design. One of the things that I really admire about him was how he took on packaging and the technical side. I'm half-designer and half-engineer, so I have a deep respect for engineers and what they do. So when I came to Chrysler, I quickly found out who Tom Gale was. What I admired about him was he was well versed in business, design, and engineering. And that's kind of how I modeled myself: that ability to speak to the various disciplines. It's one of the reasons I went to business school and got a business degree. Why I feel like an engineer because of taking cars apart my entire life and rubbing elbows with them daily. And then, of course, why I trained formally as a designer.

So I think Tom Gale was probably a very influential figure for me: Because he really taught me a lot about the balance ... Because just styling alone isn't enough ... You've got to consider the whole picture. Holistically see a vehicle. One of the things that also really impressed me about Tom was his business savvy and that he understood how to tug at the consumer's emotions. How to set up a brand using some paradigms. Going right to the heart of people with nostalgia in their minds, while also looking for white space in the industry. Because Chrysler was always kind of an underdog, we really had to understand new opportunities and white space, whether it was minivan, PT Cruiser, Pacifica, Jeeps, etc. I can go on and on—there was always a focus on opportunities to look for white space. And that was really based on understanding customer and latent needs. On

understanding the marketplace, not just putting pen to paper and drawing a cool car but first looking at the 360° picture.

**Question 2: The FCA group has a history of identifying new market opportunities and seemingly inventing segments; in effect, changing the game. Take, for example, the original minivan, or many Jeep products. In the Chrysler, Jeep, or Dodge brand history, can you describe a turning or transition point that you witnessed or were part of? And what were the resulting products that altered the brand perception?**

▶ I've been at Chrysler since 1992 and I remember seeing the minivan reinvented. In 1996, when it became sportier than it had ever been. It went on to become one of our best sellers ever. I witnessed the PT Cruiser being born—the really interesting nostalgic car that also had this combination of hyper functionality. And then I saw the Prowler being created. And that was a vehicle that really had no segment at all. It was a hot rod, it was a sports car, it was an aluminum study, and it was probably one of the more frivolous things that we've done. Pacifica came along and tried to bridge the crossover question but probably five years too early. And even the 300: mixing improbable luxury at an incredible price with American and German values fused together (see Figure 1.9).

There was a process applied to those cars that was consistent. There was always a very far-reaching concept first. There was always a very clear understanding of the opportunity to satisfy a customer: their emotional needs, functional needs,

usually some sort of combination of the two. We'd blend these things together and then go talk to people. We would show them ideas NOT to measure whether they liked it or not ... BUT rather to actually measure the tension: how and why they loved it versus how and why they hated it. It was always fascinating to me, we always used it to talk and discuss about tension, tension, tension.

Typically, when you do something (new), most people don't like it at first. Then they come to really love it so you have to be very conscious of that as a designer. You are not just doing something for yourself but you're actually trying to create and inform taste! I love the term taste-maker: fashion designers do it, moviemakers, architects, etc. It's definitely a challenge trying to convince a corporation to spend $1 billion on an idea, but designers do it as well. I'll never forget when we were working on the 300; a lot of people didn't get that car in the beginning. I got it the minute I saw it. We did research and there was always a lot of tension around the grill. Its scale and size some people even found offensive. We did the original Nassau concept and then evolved it into production (see Figure 1.10). In fact, the final grill wound up being almost as big. Which was the right thing to do because respondents had strong feelings in one way or the other about it. Some would say: "That's the ugliest car I've ever seen." Others would say: "That's the best looking car I've ever seen." There was always this strong dichotomy in the discussion.

The bosses, especially Tom Gale, totally got it. For some, it was a tough sell. Chrysler in the end always gave us that ability to

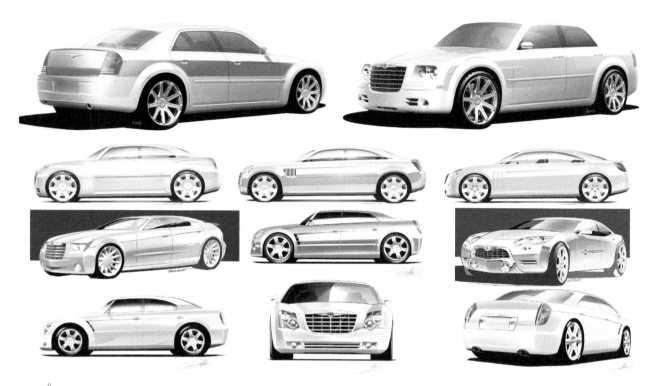

△
**1.9** Design development sketches from the team responsible for the original Chrysler 300 M
SOURCE: IMAGE CREDIT, FCA DESIGN.

△
**1.10** Chrysler Chronos concept shown on the left, the application of the theme created by the advanced design studio for the Chrysler Nassau concept executed for internal study, and the final application of the theme on the original Chrysler 300 M production car seen on the right-hand side
SOURCE: IMAGE CREDIT, FCA DESIGN.

explore and take chances because we always had to be more relevant than everyone else. We always had to have another reason to purchase our product. And that's become embedded in who I am as a designer and it's part of our DNA, to tolerate the risk in a way, but also calculate it. It has to be calculated! It can't be risk for the sake of risk. There are vehicles that literally sell in single digits because of that reason. It's not just risk; it all has to be very well calculated.

**Question 3: What types of customer needs did the design group consider in addressing those opportunities and market targets? What type of brainstorming strategies and techniques do you encourage to get the design team going at the very beginning of a project? In terms of calculating risk, what methodology do you use for understanding user wants and needs, addressing them in the design process?**

▶ We like to call them latent needs. Needs that they are unaware they have. When interviewing a frustrated minivan mom, for example, we are very indirect about it. We just let them talk.

For designers, it doesn't necessarily have to be at an organized clinic. It can even be friends at a cocktail party. I never stop working whether I'm out socially, traveling or talking to a journalist. I'm seeking data points all the time. It's funny sometimes, you can freak a person out by just going up to a lady and asking her, "Hey, how do you like your bag?" And you always sense a latent, you get a sense of stress when they talk about the styling or the way they speak about how they can't put it between the seats where there's nowhere to place it nicely in a car. Or if there's just something about the vehicle that doesn't make them feel good. It doesn't jive with who they are as a person. This is because a lot of people view cars as an extension of themselves and who they are as a person. So it's us that have to do the code matching. Connecting their personalities to what they want to achieve functionally and emotionally.

And this transfers into the brainstorming that we do. It's about also informative conversations and talking to lots of people. Good brainstorming comes from discussions. As designers, we just have to remember to inform the discussion with varied inputs. To never fall into the trap of only talking to your friends and designers. Speak with engineers, speak to many different disciplines, they all have a very unique view on vehicles. To the student, even if you're not part of an organization, sometimes it's easy just to talk to someone at Starbucks or a coffee shop on campus. And always look for the whole picture and identify the space where no one else has ... We do a lot of product positioning charts, quadrant charts as reference, and when I see these I always ask myself, why is no one in that corner? But then you have to dig deeper, and be wary of repeating someone else's mistakes.

**Question 4: Looking toward the future, while considering new technologies, what are the biggest challenges facing designers in identifying new opportunities and market segments, and what is your advice to designers to conquer those challenges?**

▶ I think that autonomy will lead to a broader spectrum of experiences and products for people. And, of course, for the foreseeable future, we will have a long period of semi-autonomy. And even within that, it will lead to different things for different people. At the moment there are more questions than answers. The only thing for sure is there will be a long period of transition. We know from our research studies that one

issue is to how to make people feel at ease when not in control. Another issue is how it will truly become a safety feature for cars beyond just accident avoidance, etc.

Because the future is unknown, the only thing that's certain is that designers will have to accommodate a broad spectrum of applications. There is a lot of research done in places like Palo Alto, and that will inform one point of view. But we also have to think about how those vehicles will be used in an unknown city in China, for example.

The bottom line is we always have to keep the emotion. Designers can never submit to the commodification of transportation. It's funny sometimes—it does keep me up at night—the idea of visual pollution and people not caring about how vehicles will look in the future. But then I think brands will still be important, and aesthetics in personal statements will still be important. It's ultimately up to designers to always remain creative!

**Notes**
1 Lovell, Sophie. *Dieter Rams: As Little Design as Possible* (London: Phaidon Press, 2011).
2 IDEO. *Design Kit: The Field Guide to Human-Centered Design.* Available at: www.IDEO.org

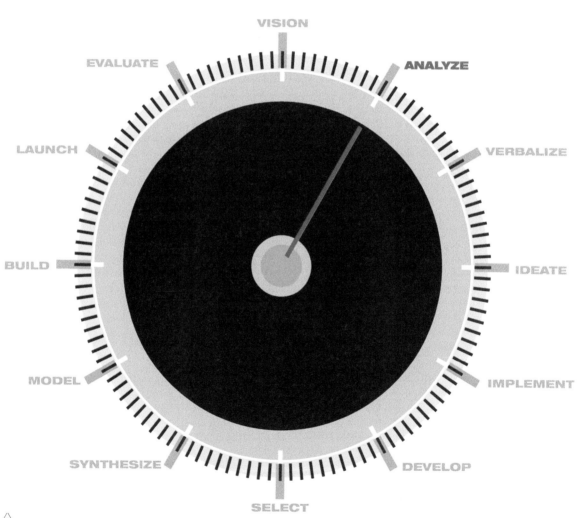

△
**2.0** Process locator gauge
SOURCE: IMAGE CREDIT, JORDAN MEADOWS.

**ANALYZE**
UNDERSTANDING A USER

CHAPTER 2

# ANALYZE

Understanding a User and Visually Showing
Why an Idea Should Be Used in a Given Experience

### Introduction to the Idea of Narrative in Design

The design process is a journey that begins with establishing a creative vision. However, before we go any further, it's crucial to "analyze and understand" a user, placing them at the center of the endeavor (see Figure 2.0). When we combine a creative vision with an understanding of a user's motivations forming a concept, we can then effectively define the *who* and *what* targets (of *where* we're headed for our journey). And though each of these is essential, the element that binds them together is the **narrative**. Narrative establishes the *why* component of the concept, and in a crucial way, is the stage where the project takes on a true human element. Having a narrative supplies motive and enables the designer to bring meaning to the exercise.

### What Specifically Is Narrative in the Design Process?

The dictionary definition defines narrative as such:[1]

> nar·ra·tive
> /ˈnerədiv/
> Noun
> Noun: narrative; plural noun: narratives
> - A spoken or written account of connected events; a story.
> - The narrated part or parts of a literary work, as distinct from dialogue.
> - The practice or art of telling stories.
> - A representation of a particular situation or process in such a way as to reflect or conform to an overarching set of aims or values

The design process uses the same definition, only with visuals. And in this regard, a narrative is the means of organizing the functional and emotional attributes of the product experience to maximize its meaning for the user.

Since we are dealing with a physical and visual medium rather than a verbal one, narrative storytelling, or **storyboarding** in the design process, is an essential step for communicating and clarifying key attributes of the design that we've laid out. Storyboarding the intended usage is a powerful tool, whether organized in a very sophisticated modern form, or communicated in basic primitives with the directness of cave paintings—this is where key elements of usage are illustrated.

Note: we're not quite ready to start with sophisticated renderings and sketches. We'll get to that fun part shortly. At this stage it's more important to concentrate the creative energy primarily on illustrating *why* the idea is used (see Figure 2.1).

The aim and focus are:

1. *To show why something is used, regarding your intended functional attributes:* Modern-day vehicles are complex. They offer a wealth of possible functions. When developing products to meet a user's needs, it is essential that we take into account the sequencing and order of operations necessary to engage with our vehicles. This holds true for a simple encounter with a door handle, to a graphical layout with an instrument panel. No matter which aspect of the vehicle you're focused on, at this stage of the process it's important to define usage scenarios. When developing a storyboard of possible functional attributes, ask yourself *why* the user would be attracted to a particular solution that you are delivering. Map out the journey visually calling out the unique selling points of your proposed experience. Graphically call out the key features that will be important to the user. Illustrating *why* your idea offers a unique functional solution goes a long way to adding resonance to every step of the journey moving forward.

2. *To show why something is used regarding emotional attributes:* As previously mentioned, this is the stage in the process where a project develops a human quality. And just as humans are pragmatic and in need of functional solutions, we also make very important decisions with our emotions. Again, storyboarding is a great tool for organizing and communicating usage scenarios. Graphically communicating *why* a proposal fulfills an emotional want or need for a user is also essential for the design process to move forward. When developing a storyboard for your intended vehicle, one of the most powerful questions you should ask is *why does this solution offer a meaningful emotional experience for the user?* Having a good answer could be the ultimate key in making the project successful! Conversely, if this basic question isn't answered properly, it won't matter how cool the thing looks. It won't matter if it's met all of the business objectives in the development process. It will ultimately be unsuccessful because it won't resonate with users. People simply won't get it: the user won't understand *why* they should choose your vehicle above other more compelling options.

△
**2.1** Graphic indicating central focus of the *why* part of the design equation
SOURCE: IMAGE CREDIT, JORDAN MEADOWS.

There's a good deal of data to conclude that in the marketplace users are motivated with their emotions rather than pragmatic goals and objectives. In this regard, focusing the storyboarding exercise on describing a means for emotional fulfillment becomes mission critical.

Many major transportation and vehicle providers understand this fact and rely on the work of psychologists to decode user wants and needs. Much has been written about the subject, and one could spend an entire career focused on this aspect of the vehicle development process. However, some over-lying basics are essential for every designer to grasp when developing a strong narrative.

In the mid-1940s, psychologist Abraham Maslow developed and popularized a theory of human motivation defined by a hierarchy of needs.[2] The basic principle placed fundamental physiological needs at the bottom of a pyramid of motivations that all people experience and understand. Working upward in the pyramid, Maslow theorized that safety, love and belonging, and esteem are needs that people would grow to desire after having met the core motivations below. Maslow also theorized that a need for self-actualization was at the top of the pyramid, being the pinnacle of needs that a person could experience after all others had been fulfilled (see Figure 2.2).

When developing a narrative for your vehicle, consider which user needs you may be fulfilling. For example, basic public transport or rudimentary mobility must offer safety and be an essential physiological service for functioning in society. Alternatively, if you are developing a vehicle that provides esteem for an individual, or communicated self-actualization, the storyboard could be composed quite differently.

Louis Cheskin, another influential psychologist and researcher, developed and popularized key theories with regard to product marketing and the impact of design on users. Bringing scientific methodologies developed through countless hours of consumer research, his consultancy (called Value Added) confirmed that not only do aesthetics and design affect perception, but the visual semantics of a product could be used to entice and speak to a user's desires.[3]

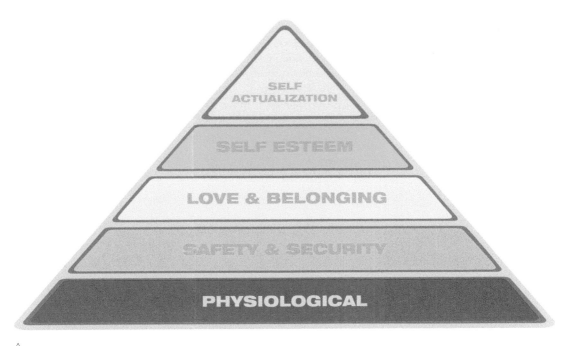

△
**2.2** Maslow's pyramid of needs
SOURCE: IMAGE CREDIT, MASLOW (1943).

These reoccurring deep-seated psychological desires appeared again and again in countless research sessions with tens of thousands of consumers, whether interviewing for what they wanted in the next Mustang or a Marlboro cigarette ad. These core desires involve a user's want for: "Accomplishment, beauty, community, creation, duty, enlightenment, freedom, harmony, justice, oneness, redemption, security, truth, validation, and wonder."[4] And while they have different levels of importance to different users, they refer to human desires that almost every consumer would and could experience at some point while selecting a product.

Considering the work of psychologists like Cheskin and Maslow in the storyboarding process allows us to create a very comprehensive list of a potential user's wants and needs. From an emotional standpoint, they range from the nice to have to the very necessary. Which ones are more important for your brand to deliver? Which ones resonate more with your targeted customer? Deconstructing and illustrating how they are delivered to your user form the function of good storyboarding. Providing a means for your product to deliver on one or a combination is crucial to your vehicle's success.

As stated before, every meaningful design exercise should place the user at the center of the endeavor. At this point in the journey you should really have a good understanding of who exactly that user is, and what their specific needs and wants are. Many organizations use different ways to codify and describe

their users. Regardless of their titles it's important to understand that different users will require a different balance of wants and needs and narrative outcomes. One way to define a user without looking at specifics such as age, or geographic location, involves describing the user with **archetypes** (see Figure 2.3).

Swiss psychiatrist **Carl Jung** was another hugely influential figure in the first half of the twentieth century, defining different traits that may dominate an individual's persona. He pioneered, developed, and popularized such concepts as the "collective unconscious" and "archetypes."[5]

Jung identified 12 key archetypes through extensive research. They included: the innocent, the everyman, the hero, the caregiver, the explorer, the outlaw, the lover, the creator, the jester, the sage, the magician, and the ruler. With regard to storytelling in the design process, it's often helpful to consider if your targeted user may have a persona that is aligned with some of these archetypes. Jung also theorized that when breaking down these archetypes into four groups of three, one could see natural orientations of these personas toward freedom versus order, and ego and the self, versus the social and group. This gives yet another layer of understanding as to which of your user's wants and needs may be prioritized in the design process.

Finally, yet another filter for understanding, deconstructing, and defining different types of users may involve looking at generational groups (see Figure 2.4). These may define overarching tendencies displayed

by individuals based on age and life experience. These groups include: Generation Y, Generation X, Baby Boomers, etc.

As you can see, there are many ways to define a user, and the most comprehensive descriptions include some sort of combination. For example: your targeted user may be a "Baby Boomer" who displays an "explorer's persona," with orientation toward "freedom" and "self-actualization." This would warrant a completely different approach from someone who is a "Generation Y," a "care-giver persona," with a "very social" orientation and a need for "love and belonging." Imagine the storyboards for those two individuals and the different wants and needs that you as a designer would have to address. Compiling one's user analysis and research is an essential part of starting a project. Shown here are examples generated by Ken Nagasaka, a Toyota designer who created a Toyota Cross Cruiser **concept car** as a student participating in my class at the ArtCenter College of Design. Ken has graciously contributed his work as reference for user profile and narrative development (see Figures 2.5–2.8). When all combined together, Figures 2.5–2.8 indicate the general direction in which to move the project, so targeted sketching can take place. Examples of how you can apply this early-stage research will be covered in later chapters of this book.

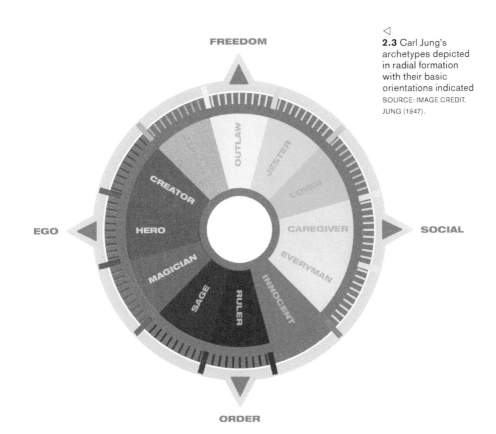

◁
**2.3** Carl Jung's archetypes depicted in radial formation with their basic orientations indicated
SOURCE: IMAGE CREDIT, JUNG (1947).

FREEDOM

OUTLAW

JESTER

CREATOR

LOVER

EGO · HERO · CAREGIVER · SOCIAL

MAGICIAN

EVERYMAN

SAGE · RULER · INNOCENT

ORDER

▽
**2.4** Typical example of generational chart outlining key groups, their formative experiences, and examples of defining products
SOURCE: IMAGE CREDIT, JORDAN MEADOWS. ADAPTED FROM PEW RESEARCH STUDIES.

| TRADITIONALIST (SILENT GENERATION) | BABY BOOMERS (BOOMERS) | GENERATION X (GEN-EXERS) | MILLENNIALS (GENERATION-Y) | GENERATION Z |
|---|---|---|---|---|
| PRE-1945 | 1946 - 1964 | 1965 - 1980 | 1981 -1997 | POST - 1997 |

**FORMATIVE EXPERIENCES:**

TRADITIONALIST:
WORLD WAR 2
THE GREAT DEPRESSION
DEFINED GENDER ROLES
NUCLEAR FAMILIES

BABY BOOMERS:
THE COLD WAR
CIVIL RIGHTS MOVEMENT
SPACE EXPLORATION
THE SEXUAL REVOLUTION
JFK ASSASSINATION
VIETNAM WAR
THE SWINGING SIXTIES
HIPPIE CULTURE

GENERATION X:
INTRODUCTION OF THE FIRST P.C.
END OF THE COLD WAR
FALL OF THE BERLIN WALL
THE RISE OF DIVORCE
LATCH KEY KIDS
EARLY MOBILE TECH

MILLENNIALS:
9/11 TERRORISM
THE RISE OF SOCIAL MEDIA
SCHOOL SHOOTINGS
IRAQ AND AFGHAN WAR
REALITY TV

GENERATION Z:
THE GREAT RECESSION
TERRORISM
THE RISE OF BIG DATA
MOBILE PHONES
CLOUD COMPUTING
CLIMATE CHANGE
GLOBALISM

DEFINING PRODUCT: THE AUTOMOBILE

DEFINING PRODUCT: TELEVISION

DEFINING PRODUCT: PERSONAL COMPUTING

DEFINING PRODUCT: SMART PHONES

TO BE DEFINED

PROJECT GOAL The goal of this project is to appeal Fuel Cell Technology as a next generation power train. This project will introduce exciting experience to consumers by using Fuel Cell technology.

FUEL CELL

FEEL RELAXED

GET EXCITED

△
**2.5** The intended emotional experience for the concept
SOURCE: IMAGE CREDIT, KEN NAGASAKA.

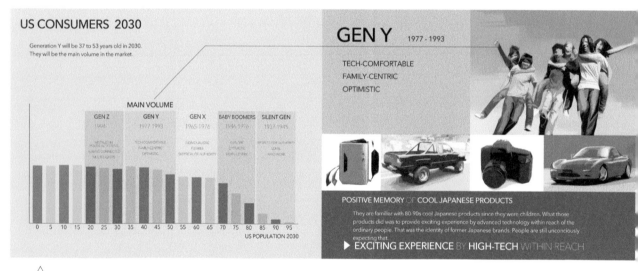

## US CONSUMERS 2030

Generation Y will be 37 to 53 years old in 2030.
They will be the main volume in the market.

MAIN VOLUME

| GEN Z | GEN Y | GEN X | BABY BOOMERS | SILENT GEN |

US POPULATION 2030

GEN Y  1977 - 1993

TECH-COMFORTABLE
FAMILY-CENTRIC
OPTIMISTIC

POSITIVE MEMORY OF COOL JAPANESE PRODUCTS

They are familiar with 80-90s cool Japanese products since they were children. What those products did was to provide exciting experience by advanced technology within reach of the ordinary people. That was the identity of former Japanese brands. People are still unconsciously expecting that.

▶ EXCITING EXPERIENCE BY HIGH-TECH WITHIN REACH

△
**2.6** A broad overview of the intended target user
SOURCE: IMAGE CREDIT, KEN NAGASAKA.

EXPERIENCE
RELAXING
INSPIRATION
BOAT TAIL PROPOTION
AERO EFFICIENCY + DECK SPACE
CATAMARAN STANCE
EXCITING

△
**2.7** An image board created to display key product influences that provide
the user the ability to enjoy a particular intended experience
SOURCE: IMAGE CREDIT, KEN NAGASAKA.

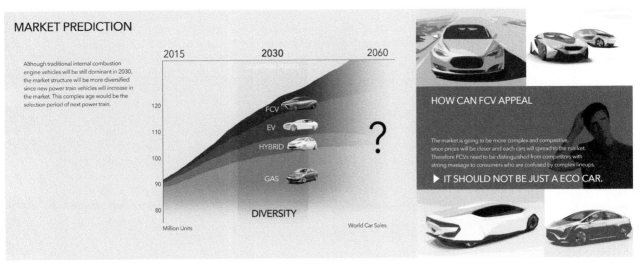

## MARKET PREDICTION

Although traditional internal combustion engine vehicles will be still dominant in 2030, the market structure will be more diversified since new power train vehicles will increase in the market. This complex age would be the selection period of next power train.

2015  2030  2060

FCV
EV
HYBRID
GAS

120
110
100
90
80

Million Units

DIVERSITY

World Car Sales

**HOW CAN FCV APPEAL**

The market is going to be more complex and competitive, since prices will be closer and each cars will spread to the market. Therefore FCVs need to be distinguished from competitors with strong massage to consumers who are confused by complex lineups.

▶ IT SHOULD NOT BE JUST A ECO CAR.

△
**2.8** Broad overview of market analysis depicting a trend toward eco-friendly, zero emissions vehicles
SOURCE: IMAGE CREDIT, KEN NAGASAKA.

### Why Is a Narrative Important?

In a world with so many different types of personalities, there's simply no clearer way to define the nuances and subtleties displayed in preferences and tastes. Added to this is the fact that humans are pack animals, hard-wired for social interaction. And as primates we evolved to have some tendencies that we just cannot seem to shed. One of them is the idea that we often anthropomorphize the things we see, feel, and experience. In short, we are psychologically prone to assign a human quality to the things we see and encounter. Psychologists believe this is a result of evolution, nature, nurture, language, and a multitude of factors. Consider how easy it is to describe the front of something as its face; so if fronts can become faces, and beginnings can become heads, then it's quite easy to understand how two lights can become eyes. All of a sudden the vehicle you are designing instantly takes on a persona and has a character whether intended or not. And as such, it has a relationship in narrative form to its user.

Narrative and storytelling are also very important in that we as humans have a natural curiosity about us. Every child grows up asking why? *Why* is how we inform ourselves and demystify the world around us. Storytelling is an inherent part of education and teaching. It's how we relate experiences as individuals to one another. It's how we share common bonds as a group.

Author, scholar, and mythologist Joseph Campbell dedicated his life to helping the world understand the functions of storytelling and myth. He theorized that we as a species have come to rely on storytelling and myth as a means for survival in four key ways. In this regard, stories serve one function: a metaphysical function, they help us to explain the mysteries of life and reoccurring symbols. In a sense this could form the basis of religion. Second: stories as passed down through the ages have acted as a form of pseudoscience and served to help humans understand the Earth, nature, and reoccurring phenomenon such as the seasons, tides and astrological functions, etc. Third: stories are key to sharing experiences, education, learning, and serve as a means for depicting ideals. One can hardly imagine going through school at any level without relaying a fable, tale, or anecdote. And, finally, stories and narratives serve a sociological function, bringing structure and order to our communities, tribes, and governing bodies.[6]

With this in mind we can easily see how crucial stories and narratives are to us as human beings. And with so many different types of personalities, and the various combinations and opportunities to perceive objects with personality, storytelling becomes the only way to establish and ensure patterns of appreciation and understanding. Furthermore, psychologists have also concluded that humans are wired to remember stories more easily than numbers, equations, and figures. We inherently process them with our hearts and emotions, giving them a lasting impact. This makes the idea of storytelling all the more relevant when dealing with a product with complex extensive features, such as a vehicle. For example, one

may not be able to remember the horsepower figure for their favorite sports car, but one can definitely remember the feeling of exhilaration and sense of empowerment that it gave.

Finally, and of particular relevance for us as designers, is the consideration that we exist in large corporate organizations. Vehicles require large teams to produce them. Key team members come and go over the long process of product development in the transportation world. In organizing

and motivating a team around a particular idea, it's important to remember that human beings are also wired to rally behind a cohesive and heart-moving story. The simple fact is it is just easier to get people to work collectively when following a narrative. Belonging to a group with a shared vision and goal is natural to us, and storylines help us to communicate and understand these goals. It all comes down to the simple idea that vehicle creators need to follow a script so the users can enjoy it!

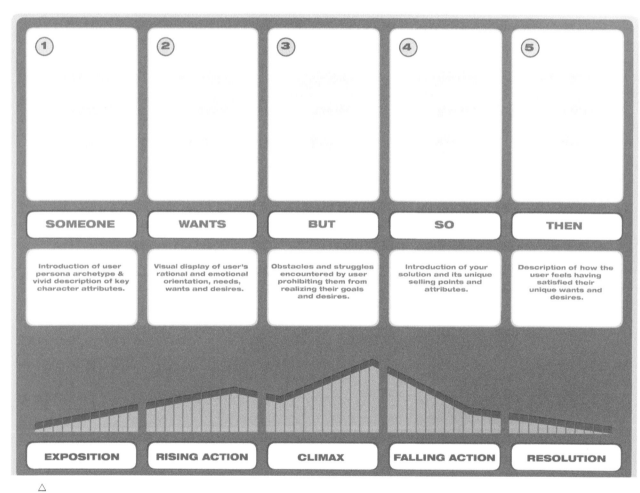

△
**2.9** A basic storyboard template outlining the compositional elements of a narrative, adapted from various open-source storytelling guides
SOURCE: IMAGE CREDIT, JORDAN MEADOWS.

ANALYZE
UNDERSTANDING A USER

## How to Construct a Narrative to Establish Motivation

Storylines and narrative almost always follow a prescribed pattern. Using an established pattern has significant advantages in that both the storyteller and user/listener inherently understand the desired outcome. This gives both parties the opportunity for a shared bond, thus giving meaning to an experience irrespective of user personality type. So in a sense it doesn't matter who you're designing for: if the subject/user/protagonist of the story is appropriately matched and aligned with the outcome of the narrative, then the product is sure to deliver a meaningful experience.

A common recurrent narrative structure can be deconstructed into five phases. When outlining a storyboard for the sake of convenience, one can illustrate the narrative in these five sections. Initially there's an exposition or description of the user. The next section involves a rising action where the desired activity is developed. The climax of the story then displays the issue or problem that needs to be dealt with. The falling action introduces your product solution. And finally, the resolution depicts the desired outcome for the user and indicates how your product has fulfilled the necessary emotional and functional wants and needs. It's a common pattern that first shows Step one (somebody). Step two (wants). Step three (but can't have because). Step four (so you provided them with). Step five (then they can do___ and feel___) ... (see Figure 2.9).

*1 Somebody. / 2 Wants. / 3 But. / 4 So. / 5 Then.*
This is a common five-phase structure that works in design and not coincidentally is derived from a universal narrative pattern dating back to the Dark Ages: Exposition/rising action/climax/falling action/resolution.

An even more succinct structure can be broken down into three phases that work on the same idea. For example, Step one: your user as you defined them experiences a want or need. Step two: they are confronted with an obstacle or difficulty in fulfilling that want or need. Step three: your product is introduced to them and gives them satisfaction. This is a very common three-phase structure that is composed around the idea of (1) separation; (2) transition; and (3) reintegration.

To further add relevance for your vehicle and align it with your **mission statement** and brand, you can deliver this narrative in the form of an established plot. Most of these are also pre-existing and commonly understood, giving you another chance to develop a strong connection between the vehicle's intended usage and its user. In storytelling, there are seven basic plots:

1. Overcoming a monster
2. A rags to riches scenario
3. A quest
4. The voyage and return
5. A farce or comedy
6. A tragedy
7. A tale of rebirth.[7]

So just imagine, for example, that female user that we earlier defined as having an "explorer's persona," with orientation toward "freedom" and "self-actualization" would really appreciate a simple three-phase

structure delivered in a "quest" or "voyage and return" scenario. With just that alone, we can imagine the type of vehicle we would design for that person. We can even imagine how they would use it. And we can definitely imagine the type of brand it might have. And with these few well-organized components we can then start to give shape and form to this concept. If required, we can even be more specific and outline the basic **UX** "user experience" in the form of a journey map (see Figure 2.10).

The use of these various concept building tools is essential in constructing the framework for a user experience exercise. The application of these principles will be explored later in the book as we begin to dive into design execution. Whether it's in the form of journey mapping or communicated in high-level storyboards like the two examples from Ken Nagasaka's Cross Cruiser concept, an analysis and understanding of the vehicle's intended usage is the key objective at this point (see Figures 2.11 and 2.12).

In summary, narrative is the key to loading a user experience with emotional meaning. It establishes the motive and forms the *why* component of the design. Using pre-existing structures allows for infinite combinations of user types to understand and visually interpret how their wants and needs can be met by the product.

When the designer has an understanding of the user, their wants and needs, and can develop a strong narrative, they are empowered with the tools to place that individual in the position of the protagonist or hero. When this occurs, the user is certain to have a meaningful experience.

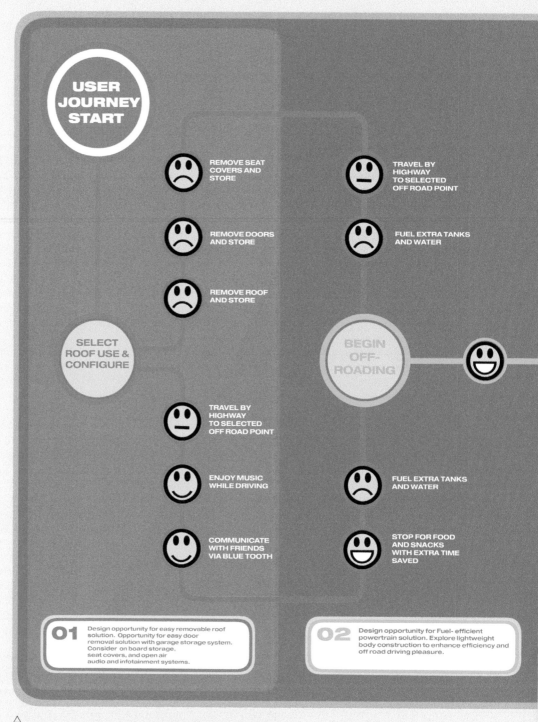

USER JOURNEY START

REMOVE SEAT COVERS AND STORE

REMOVE DOORS AND STORE

REMOVE ROOF AND STORE

SELECT ROOF USE & CONFIGURE

TRAVEL BY HIGHWAY TO SELECTED OFF ROAD POINT

ENJOY MUSIC WHILE DRIVING

COMMUNICATE WITH FRIENDS VIA BLUE TOOTH

TRAVEL BY HIGHWAY TO SELECTED OFF ROAD POINT

FUEL EXTRA TANKS AND WATER

BEGIN OFF-ROADING

FUEL EXTRA TANKS AND WATER

STOP FOR FOOD AND SNACKS WITH EXTRA TIME SAVED

01 Design opportunity for easy removable roof solution. Opportunity for easy door removal solution with garage storage system. Consider on board storage, seat covers, and open air audio and infotainment systems.

02 Design opportunity for Fuel- efficient powertrain solution. Explore lightweight body construction to enhance efficiency and off road driving pleasure.

△ **2.10** A typical journey map, a more detailed way of developing the narrative in conjunction with a storyboard, the image here depicts off-road outing as an example. Journey maps are essential in outlining key touch points and providing emotional analysis.
SOURCE: IMAGE CREDIT, JORDAN MEADOWS.

**ANALYZE**
UNDERSTANDING A USER

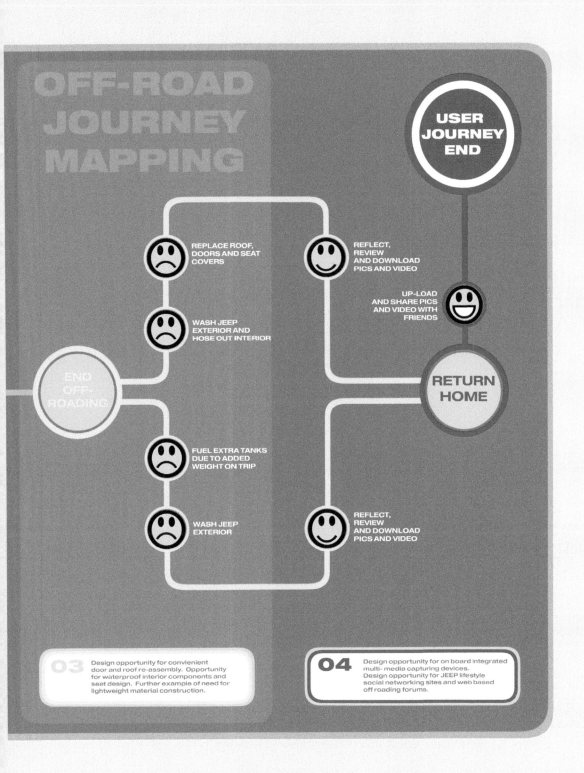

OFF-ROAD JOURNEY MAPPING

USER JOURNEY END

REPLACE ROOF, DOORS AND SEAT COVERS

REFLECT, REVIEW AND DOWNLOAD PICS AND VIDEO

UP-LOAD AND SHARE PICS AND VIDEO WITH FRIENDS

WASH JEEP EXTERIOR AND HOSE OUT INTERIOR

END OFF-ROADING

RETURN HOME

FUEL EXTRA TANKS DUE TO ADDED WEIGHT ON TRIP

WASH JEEP EXTERIOR

REFLECT, REVIEW AND DOWNLOAD PICS AND VIDEO

**03** Design opportunity for convienient door and roof re-assembly. Opportunity for waterproof interior components and seat design. Further example of need for lightweight material construction.

**04** Design opportunity for on board integrated multi-media capturing devices. Design opportunity for JEEP lifestyle social networking sites and web based off roading forums.

## GEN Y Family

| 01 Weekdays 7 AM | 02 Weekdays 11 AM | 03 Weekdays 5 PM | 04 Weekdays 5 PM |

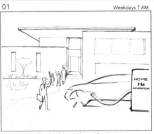   

GEN Y Family : 42 male, 40 femal, 13 boy, 10 girl
Their home is in suburban area

A man is leaving home to work in the morning.
A car is charged by a home H2 generator.

He is working for a insurance company.
The office is in urban area.

He often goes to meet clients by his car.
He wants his car to be not too casual.

Commute on congested highway everyday.

He is so frustrated with the situation everyday
and wants to escape from there.

While commuting, he imagines his dream vacation.
But he knows it costs too much for him.

## GEN Y Family

| 05 Weekends 7 AM | 06 Weekends 12 AM | 07 Weekends 1 PM | 08 Weekends 4 PM |

He is preparing for recreation with his family on weekends.
The grass hatch can slide to open for enough cargo space.
FCV system can charge electric recreational mobilities.

They arrive at desert area.
Start playing with electric recreational mobilities.

They arrive at desert area, and play with "Towing Mountain Board"
just like a wakeboard.
The cargo space can be a deck for passengers.

Later they start fishing by a lake, reading a book, or whatever
for refreshment.

## GEN Y Single

| 01 | 02 Weekdays 8 AM | 03 Weekdays 6 PM | 04 Weekdays 7 PM |

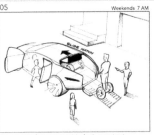   

GEN Y Single : 38 male
Living in urban area

He is working for an IT company.
The office is in urban area.

He always works on computer screens.
He wants to go out to refresh after work or weekends.

He picks up his girl friend on the street near his office for a dinner.

He drives an ocean side highway to show a sunset.

He wants driving performance and comfortability on his car.

## GEN Y Single

| 05 Weekdays 8 PM | 06 Weekends 8 AM | 07 Weekends 1 PM | 08 Weekends 4 PM |

 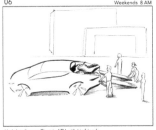  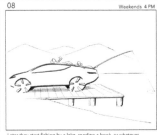

On the way back his home, he buys a Christmas tree.
He opens the hatch and loads the tree on the space smartly.

He is loading an Electric ATV with his friends.

They arrive at desert area, and play with "Towing Mountain Board"
just like a wakeboard.
The cargo space can be a deck for passengers.

Later they start fishing by a lake, reading a book, or whatever
for refreshment.

△△
**2.11** Storyboards for Toyota Cross
Cruiser concept, depicting usage
scenarios for Gen Y families
SOURCE: IMAGE CREDIT, KEN NAGASAKA.

△
**2.12** Storyboards for Toyota Cross
Cruiser concept, depicting usage
scenarios for Gen Y singles
SOURCE: IMAGE CREDIT, KEN NAGASAKA.

**ANALYZE**
UNDERSTANDING A USER

## Q&A

**ANGELA WELTMAN**
ON CONSUMER INSIGHTS

*A cognitive psychologist, Dr. Angela Weltman specializes in examining underlying motivations of behavior to inspire automotive product design. Much of her work focuses on the deep-seated emotional triggers tied to archetypal goals, and the relevance of metaphors and storytelling in consumer understanding of objects and new design. By marrying academic theories (Jung, Freud, cognitive science, etc.) with social and demographic/generational trends, Dr. Weltman finds creative solutions to product development and innovation. Over the last 24 years, Dr. Weltman has worked with most of the major automotive manufacturers. Dr. Weltman received her B.A. from UC Berkeley in English Literature and her Ph.D. from UCLA in Cognitive Psychology.*

*She has graciously agreed to share some of her insights on consumer behavior and the importance of narrative in design (see Figure 2.13).*

△
**2.13** Angela Weltman, cognitive psychologist
SOURCE: IMAGE CREDIT, ANGELA WELTMAN.

**Question 1: Who are your personal professional heroes? Who do you most admire, and why? How does this key influence relate to, or impact the work you do for automotive and transportation companies?**

▶ The people that I think about and use as a North Star in my work are people who brilliantly break barriers and help us see or understand the world in a new way.

Carl Jung.[8] He was a Swiss psychiatrist and psychotherapist who broke boundaries in the way he looked at and studied the world. Jung was very introspective, and appreciated and embraced the logic of science as well as the emotional and spiritual elements of our shared humanity. He risked searching deep into the depths of his own psyche and, in the end, illuminated an incredible way to think about identity and personal narratives based upon archetypes and the journey toward individuation. He also developed ideas around the eternal connections we have to each other through a collective unconscious. In essence, Jung helped us think about our lives in terms of a mythic journey. A life path that requires overcoming obstacles to progress toward our most true and ideal self as defined by a deep archetypal construct. In modern times, Jungian archetypes often go by names such as Hero, Jester, Sage, Lover, Rebel … They guide a person on his or her mythic journey and reflect the deep-seated motivations and goals of one's ideal sense of self.

My team and I use Jung's ideas about archetypes and striving toward an ideal self in our work. We study *who people are* as well as *who they want to be.* We study the deep-seated motivations and goals tied to a target customer's archetype ideals, and examine how these motivations and goals manifest themselves in relevant and compelling design. Over and over again, we find people love design that communicates the deep-seated motivations of the archetype. For example, the Rebel archetype feels passionate about design that communicates the rebel's ideals: aggressive, tough, set against the status quo; the Lover archetype loves sensual design, mysterious, warm and inviting.

People feel passionate about design that engages the deeper, emotional, motivations and goals of the archetypal self because it *helps people become more of who they want to be*, and *reaffirms the ideals of the aspirational self.*

Jung also inspires me to think beyond the confines of a rational,

logical approach to my work. I am a scientist and I conduct research. The rules can be pretty cut and dry. But the truth is poetic. Truth requires art and science. To understand people and their needs and desires revolving around design, one must be disciplined as well as creative, and open to the emotional underpinnings of our humanity. And one must make it personal. Like Jung did.

Other influences include Daniel Kahneman and Amos Tversky, two renowned psychologists, who studied judgment and decision-making.[9] Together they developed breakthrough experiments to examine how people think and found that people do not employ rational rules of problem-solving, but instead use heuristics, cognitive shortcuts, in their reasoning. Kahneman and Tversky's approach and findings revolutionized the field of Cognitive Psychology and Economics. Not only did they uncover biases in our thought processes, but they created brilliant methodologies that allowed us to "see" how the mind worked, to see the unseen.

In my work, I often think of Kahneman and Tversky, and respect people's seemingly idiosyncratic attitudes and behavior that come from cognitive heuristics. For example, I understand that people's decision-making and choices can depend on context and framing of the situation, and so I'm careful in setting up a context for design, and couch my analyses appropriately. Kahneman and Tversky also inspired me with their clever research methodologies. To understand how people perceive

objects and design, we need to understand elements that can be seen or easily measured, elements such as space, horsepower, value/price. But we also need to understand elements that cannot be seen or easily measured, like the meaning and communication of a design, or underlying emotional triggers. Kahneman and Tversky taught me to pursue new approaches to "see" the unseen. I lean heavily on metaphorical language and reasoning (thank you George Lakoff, a masterful cognitive linguist), storytelling and abstract visual stimuli (my own extension of a type of Rorschach test).[10]

Which leads me to a couple of other influences: Antonio Damasio[11] and Joseph LeDoux,[12] two neuroscientists who revealed the importance of emotion in cognitive processing and decision-making. They show that personally satisfying decisions are largely dependent on emotions. So to understand if a design is relevant and compelling, it's imperative that we understand the emotional underpinning of a design, and uncover the emotional narrative revealed in the archetype.

**Question 2: Why are stories and narratives so effective in motivating individuals and groups to act, use, or purchase one vehicle versus another?**
▸ Not all stories and narratives motivate people. Narratives are powerful, they motivate people to act, use or purchase a vehicle, when they bring to life the deep-seated archetypal journey, and connect the vehicle to a target customer's aspirational motivations and goals. These deep emotional narratives allow people to envision the dream of what they

want to become. When the vehicle becomes tied to this dream, the vehicle becomes a compelling conduit that guides, realizes or reaffirms the ideals in the mind of the customer. In this light, the vehicle (as conduit) becomes hard to resist.

A great example: for a long time, the cowboy narrative for trucks spoke to the archetypal aspirations of rugged individualists. The narrative brought to life the cowboy myth and tied it explicitly to the pickup truck. So when someone bought the truck, they owned the powerful story too. They became more of who they wanted to be, rugged individualists. The truck, reinforced by the narrative, realized and reaffirmed the ideal self.

Whether these narratives focus on mythic quests or concrete user experiences of the archetype, all relevant stories speak to the aspirational self-ideals, and, with the vehicle as a key component, promise one's progress toward self-actualization. The narratives, and the vehicle, inspire.

About 10 years ago, I spoke to a very successful Los Angeles producer about her thoughts on luxury vehicles. She had very differentiating taste, a clear and beautiful **aesthetic**, and loved to express herself. She designed her own home in the hills of Hollywood and meticulously decorated each room. She dressed elegantly. Collected art. When I asked her about her car, she said "nothing inspired" her. And then she explained that no luxury vehicle at the time "told the story" of how she got to where she was in her life. Essentially, nothing reaffirmed the ideals of her archetype, a

narrative based on the aspirations of the Ruler-Lover. Her success came from being a warm and generous leader, from developing deep relationships with those she worked with, and from her unique appreciation of design. She felt the narratives in luxury cars focused on winner-take-all conflict and aggression of the Ruler-Warrior, and left her unfulfilled, and unwilling to buy a vehicle. *Uninspired.*

Narratives that weave together archetypal stories with the vehicle speak to those who *aspire to the archetype ideals* as well as those who *appreciate the archetype.* Ford Mustang drivers embrace the Rebel archetype and love that the vehicle narrative communicates masculine cool and breaking free; fast, confrontational, and romantic. Rebels passionately cling to the Mustang narrative. But people who *appreciate* the Rebel narrative are also drawn to the vehicle. These people like the idea of the rebel and choose the vehicle so the story, experience, image can be part of their life. Or they might be the ones who keep a Mustang in their garage and take it out on sunny Sundays or for a weekend get-away. They covet the dream, even if they don't live it.

Finally, for a narrative to be relevant and compelling, it must also be true and coherent. The pickup truck and Mustang are tied to relevant and compelling narratives because the story and vehicle together represent an honest and authentic expression of the archetype ideals.

One last comment, there are also narratives that serve as warnings, narratives that are based on archetypal fears. These narratives

are also powerful, but typically dark. In the world of vehicles, these narratives may create too much aversion and lose their ability to seduce.

Question 3: **Why do narrative structures and plot patterns recur around the world and through the ages, and are they universal? Is there one more common to vehicles and transportation?**

▶ To answer this question, I refer back to Carl Jung and his theories about a collective unconscious. Jung developed the idea that all humans, through evolution of the species, are born with instinctual knowledge of the fundamental, and eternal, archetypes. He called this shared knowledge, the collective unconscious. And he argued that this inherited and shared understanding of the archetypes has an important influence on our psychic life. For example, a newborn baby instinctively bonds with his or her mother, because of an *a priori* recognition of the nurturing Mother archetype. And conversely, at any age, we intuitively know to be wary of the Dark Knight even before we might ever meet this character. We see evidence of the collective unconscious when we examine the narratives and characters that show up throughout time and throughout the world, and find that the same narratives and characters keep reappearing. Scholars in fields such as psychology, sociology, anthropology, and even neuroscience, have found that the same archetypal ideals appear over and over again in our symbols, myths, religions, dreams, and fantasies. And so I believe that we see the same narratives and characters throughout time and throughout the world because they all spring from the innate and instinctual, eternal and universal, knowledge of the archetypes, from that deep well of the collective unconscious.

Now you ask, is there one narrative or character more common to vehicles and transportation?

From my experience, I would say no. There is not one narrative or character more common to vehicles and transportation. Joseph Campbell's Hero's Journey (see *The Hero with a Thousand Faces*, 1949), overcoming obstacles to transcend, applies to vehicles and transportation, but it applies to everything else too.

But I do think there exist three common tensions related to vehicles and transportation, underlying the archetypal narratives and characters. These three common tensions revolve around energy:

**(1) psychic energy: aggressive/ death (Thanatos) vs. sexual/love (Eros).**

Do the archetype ideals embody the need for aggression, communicated/celebrated by vehicles/transportation design that's more angular, imposing, and robust? Or, does the archetype ideal embrace sensuality, communicated/celebrated by design that's more curvaceous, inviting, and fluid? Or could there be a dance between Thanatos and Eros? These fundamental questions around psychic energy need to be answered.

**(2) spiritual energy: dark vs. light**

Spiritual energy reflects the moral standing of the archetypal ideals, and is expressed by the metaphor around dark vs. light. So we have our darkest instincts, here it's all Freudian Id, represented by design centered on pleasure-seeking appetites— cool, thrilling, explosive, naughty. Or we have a quest for light, the Freudian Superego, design that clings to the collective sense of right and wrong—earnest, bound to others, civilized. Vehicles/ transportation design can belong to the dark or light world, or they can represent a dance between our dark and light impulses. Light coming through dark for those who crave redemption or uplifting transformations. Dark coming through light for those who want to dive into secret indulgences, etc. The fundamental calibration around spiritual energy needs to be defined.

**(3) physical energy: chaos vs. order**

Do the archetypal ideals center on high-stimulation, free-wheeling chaos communicated in design that's surprising and unpredictable? Or does the archetype ideal appreciate the calm and control of order, represented in quiet design that's straightforward and highly organized? Or is there a dance between chaos and order? For example, breaking out of a controlling world, rebelling against the constraints of structured convention; or creating order in chaos, the hero's ultimate act of overcoming.

These three types of energy are the tension points that create life in the machine and the emotional triggers that connect to people. They are common to vehicles and

transportation partly because I think we believe, on some level, vehicles are alive. As young children, we see vehicles on the road, moving around, seemingly on their own. Independent movement is often attributed to an animal that is alive. I think on some subconscious level we continue to believe, from our early experience with vehicles, that they are indeed an animal that is alive.

Furthermore, these tensions, because they represent a life force, are fundamental to creating that relevant and compelling story where both the narrative and the vehicle come together in an honest and authentic expression of the archetypal ideals.

**Question 4: Looking toward the future while considering zero emissions and autonomous technology, what are the biggest challenges facing designers in delivering emotionally meaningful experiences to users? What is your advice to designers to conquer those challenges?**

▶ The biggest challenge is having clarity. My advice is *have clarity*. To move forward into the future, and deliver emotionally meaningful experiences to users, designers must have clarity about:

- *Context*: the world we live in.
- *Customer*: the deep-seated motivations and goals of the target customer.
- *Concept*: what designers are actually designing.

Sounds easy. It's very hard.

First, *context*. To have clarity about the world we live in means seeing the world of today, unencumbered by the filters of what we think we know, or of our memories of the past. If we can see the world of today as if it were the first day we ever arrived on Earth, we will face forward into the future and know the true context of design. "I'm not trying to predict the future. I'm trying to let us see the present," said William Gibson, science fiction author, in the 1970s, long before the ubiquity of the internet in the 1990s, who predicted future worlds, and defined key language, around computer networks— he first introduced the term "cyberspace."[13]

Suggestions on how to have clarity about context: look and listen; read everything you can. You must be constantly curious. Keep tons of hypotheses alive in your head, and then let everything go and then let yourself be surprised at what you see and hear. Wonder.

Next, *customer*. I have discussed the importance of the deep-seated motivations and goals of the target customer, represented by the archetypes. When thinking about moving forward into new territory, clarity of deep-seated motivations and goals becomes imperative for the motivations and goals of the target customer to direct the target's path forward— and thus direct us forward. They are the blueprint defining where the customer wants to go as they aim to realize the ideal self. In a very real sense, the future lives in the mind of the customer, in this aspirational path. We see where the customer wants to go, and so these deep-seated motivations and goals lead us to where we need to go.

Throughout the design process, it's also important to have the customer evaluate whether or not

the developing forward-looking design does indeed represent an honest and authentic expression of the archetypal ideals. In the mind of the customer, are we bringing to life the deep-seated motivations and goals in the design?

But take note: *The customer can never tell us what to design, what the vehicle or experience should be. Customers are not designers and so they should never be put in the position of designing the vehicle or experience.* People can only reveal to us their dreams of the aspirational self; it's the deep-seated motivations and goals underlying these dreams that guide the designer's vision.

Imagine this example in the world of mobile phone design. A target customer embodies the Magician archetype and has a fundamental and overriding aspiration, to continually change and transform, to shape shift by living without friction, living with flexible, versatile, fluid ease. If you were designing a phone for that customer in the early 2000s, and used *continually change and transform, to shape shift by living without friction, living with flexible, versatile, fluid ease,* you couldn't offer a compelling solution with simply more buttons. The Magician archetype requires sorcery, and this is exactly the type of solution an unforgettable group of innovators gave us with a wafer-thin screen that performs like magic.

The customer would never be able to tell a designer to get rid of the buttons on the old mobile phones, but a customer's deep-seated motivations and goals could guide—inspire and prioritize—the designer's vision of something so

revolutionary. As it turns out, in the example above, those innovative designers and engineers were true Magicians as well.

*Concept.* Clarity of the design concept comes from clarity of context and customer. Clarity of the design concept means designers know what they are creating, in the context of the new world and in response to customers' aspirations. A design concept should be specific, prioritize what's important and what's in and out of bounds. A design concept is often best when it's metaphorical.

Here's an example of clarity around design concept. Note: this is a fictional example, and is not meant to describe a real concept.

- *Context*: The vehicle used to take us out to places. With our vehicles, we moved along the highway and discovered the world. This discovery was exciting and helped us fight off boredom in our lives. In the context of the new world, everything moves toward us. We are now standing still, and everything comes at us. It feels daunting, and exhausting. The Earth even feels exhausted by all this activity. A tornado of activity. In this context of the new world, there's an overwhelming desire to escape. There's a lot in the news about other planets, the possibility of leaving the Earth and colonizing somewhere else.
- *Customer*: In this example, we define the customer deep-seated motivations and goals revolving around the Innocent-Lover. This customer finds power in a fiercely earnest desire to pivot toward the

positive and beautiful natural forces on Earth. In the context of the modern world, this group wants to break away from the gravitational pull of the chaotic tornado and fly toward: an open field of flowers, a tall forest of redwoods, white sandy beaches. Through an optimistic and relentless belief that humans and nature, together, can solve all problems, this group flies forward into the future, to save Earth and celebrate all its glory. This is a group that appreciates flights of fancy and the promise of a bright future, and avoids the overwhelming burden of fear and anxiety. This group would never abandon the splendor of Earth for a life on another planet.

- *Concept*: Butterfly. A new-energy vehicle that must be powerful enough to break away from the gravitational pull of the oppressive society while expressing a deep commitment to positive and beautiful natural forces on Earth (implications for innovative materials). Like a butterfly, the design reflects bold sensual forms, it's bright and light, dynamic, flits about. This Butterfly pivots toward the positive. Dynamically takes advantage of the weather, or the views. It filters and organizes positive experiences and information coming at the target. It communicates a renewed belief in the good with light coming through darkness and creates a very quiet and disciplined sense of order to combat the chaos of the modern world. It streamlines the Innocent-Lover's life, and dazzles the target with a sense of optimism about human potential and natural resilience.

It's a very exciting time for vehicles and transportation. In this example, we stay in the world of vehicles. But as we all know, it's important to stay open to what the concept really might be. Do autonomous vehicles morph into robots? Roving rooms? Clarity is key.

**Notes**

1 Merriam-Webster. *Dictionary*. Available at www.merriam-webster.com
2 Maslow, Abraham. A theory of human motivation. *Psychological Review*, 50(4) (1943): 370–96.
3 Cheskin, Louis. *Why People Buy: Motivation Research and Its Successful Application* (New York: Ig Publishing, Incorporated, 1959).
4 Diller, Steve, Shedroff, Nathan and Rhea, Darrel. *Making Meaning: How Successful Businesses Deliver Meaningful Customer Experiences* (Berkeley, CA: New Riders, 2005).
5 Jung, Carl. *The Archetypes and the Collective Unconscious (Jung's Collected Works, vol. 9a)* (Princeton, NJ: Princeton University Press, 1959).
6 Campbell, Joseph. *The Hero with a Thousand Faces* (San Francisco: New World Library, 2008).
7 Booker, Christopher. *The Seven Basic Plots: Why We Tell Stories* (London: A&C Black, 2004).
8 Jung, Carl. *Man and His Symbols* (New York: Dell Publishing, 1968).
9 Kahneman, Daniel, Slovic, Paul and Tversky, Amos. *Judgment under Uncertainty: Heuristics and Biases* (Cambridge: Cambridge University Press, 1st edn, 1982).
10 Lakoff, George and Johnson, Mark. *Metaphors We Live By* (Chicago: University of Chicago Press, 2nd edn, 2003).
11 Damasio, Antonio. *Descartes' Error: Emotion, Reason and the Human Brain* (New York: Quill, 2000).
12 LeDoux, Joseph. *The Emotional Brain: The Mysterious Underpinnings of Emotional Life* (New York: Simon & Schuster, 1998).
13 Gibson, William. *Neuromancer* (New York: HarperCollins, 2011).

**ANALYZE**
UNDERSTANDING A USER

EXPLORER

CREATOR

HERO

MAGICIAN

SAGE

RULER

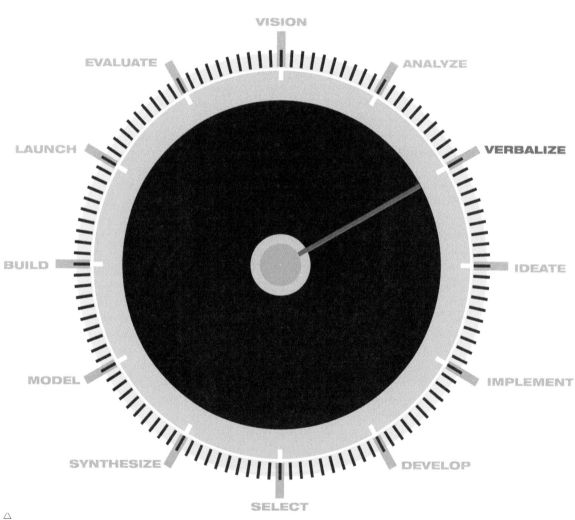

△
**3.0** Process locator gauge
SOURCE: IMAGE CREDIT, JORDAN MEADOWS.

# CHAPTER 3

# VERBALIZE
## Spreading the Word and Generating a Mission Statement

**Spreading the Word and Generating a Mission Statement**
If the initial steps of a creative journey are about vision, and dreaming of an ideal design solution to fill a user's desires, the next phase is about clarifying that vision and assembling the raw materials and conceptual components to create a design brief to "verbalize" the vision (see Figure 3.0).

In the previous chapters we discussed various methodologies for verbalizing, clarifying the vision, and getting started in the process. The main component was an understanding of a potential user. In professional settings a user can be given to you. In academic settings, students can choose or create a potential user. In the later example, selecting a profile that is interesting and engaging is essential, as the brief will follow suit. If the user profile is predictable and mundane, you'll have that much more of a challenge creating an interesting brief to suit that user.

In either case, it's your creative vision and the empathy you have for that user that will set the groundwork. Along with these two components, we also talked about a need for situational awareness of the space that you're entering (getting a feel for what's been done previously and what's going on currently), and finally we spoke about the need to source visual inspiration and influence. These key elements can then be combined in a way that suits your personal goal or organizational objectives. Assembling them in various ratios can lead to different results; for example, if a project is very much about your own creative vision, the design can become more of an artistic or conceptual exercise. If it's defined by being strongly rooted in an existing market inspiration and influence, it may come across as contemporary and stylish. If very much oriented toward a past movement or individual, it may be seen as an homage or evolutionary. Very easily one can already see how composing an understanding of user needs, creative vision, situational awareness, and visual inspiration can yield different types of framework for your project. So it's important to get the mix right and aligned with your personal goals or organizational objectives.

Whether these goals and objectives are driven by an individual designer or by a collective group, ultimately they should be reflective of a **mission statement**. Mission statements in a corporate context are used to succinctly communicate the core attributes for the brand. These typically include a set of cultural beliefs and behaviors with regard to its employees. A description of a value equation for its customers and shareholders is needed. And, finally, an indication of the role played in a broader socioeconomic and environmental context (see Figure 3.1). For example, Daimler's mission statement reads as follows:[1]

> As the inventor of the automobile, we believe it is our mission and our duty to shape the future of mobility in a safe and sustainable manner – with trendsetting technologies, outstanding products and made-to-measure services.

Some corporations use vision statements alongside or instead of mission statements. Mazda's corporate vision statement is as follows:[2]

> We love cars and want people to enjoy fulfilling lives through cars. We envision cars existing sustainably with the earth and society, and we will continue to tackle challenges with creative ideas.
> 1. Brighten people's lives through car ownership.
> 2. Offer cars that are sustainable with the earth and society to more people.
> 3. Embrace challenges and seek to master the Doh ("Way" or "Path") of creativity.

Furthermore, in some cases, corporations will define their brand goals as well. Mazda's statement of brand essence reads as follows:

> **Celebrate Driving**
>
> Mazda's Brand Essence is "Celebrate Driving." "Celebrate Driving" delivered by Mazda is not just about driving performance. Choosing a Mazda prizes the owner with confidence and pride. Driving a Mazda and developing the urge to take on new challenges. Not just our products but every encounter with Mazda evokes the emotion of motion and makes customers' hearts beat with excitement. All of these are contained in our brand essence of "Celebrate Driving."

Mission statements, corporate visions and statements of brand essence/values can be used in combination with the specific business goals and objectives of a program to create a design brief.

### Understanding the Interplay Between Brand and Design Brief

Branding is a study unto itself that every designer should have a feel for. For that matter, we can all think of examples of strong brands that we appreciate and are drawn to. Which are your favorites in the world of transportation and why? Likewise, we can easily recall brands that have done quite a poor job of managing their image and leave a lot to be desired. Consider why some go through highs and lows and some just lose their mojo and never get it back. This occurs because they either fail to deliver the promise made to their users, or adjust to meet their evolving needs.

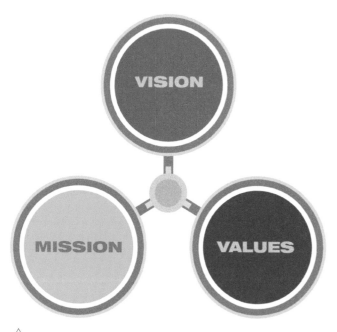

△
**3.1** Three key elements essential in developing a brand philosophy
SOURCE: IMAGE CREDIT, JORDAN MEADOWS, ADAPTED FROM OPEN SOURCE INFO.

The degree to which an existing brand heritage and DNA is leveraged when generating a **design brief** is a result of the relationship of three key factors:

1. The design brief's effectiveness in meeting a business objective.
2. The business objective's alignment with the brand.
3. The brand's fidelity in honoring its mission statement.

Managing these three components is an ongoing process. Sometimes brands drift off message. Remaining competitive requires constant focus on core users and a willingness to adjust to their evolving wants and needs. The bottom line is, the user has to trust a brand. Trust is the foundation for any relationship. Creating and delivering products that honor a brand's promise is the main way that user confidence and trust are maintained.

From an individual's perspective it's not always essential to begin a program with a specific badge or brand in mind. With student design projects, and with independent passion projects, the key factor is having a personal vision and strong point of view in lieu of a corporate mission statement. In this case you are your own brand in a sense. To create a design brief, one needs to be able to clearly state: what you believe in and stand for; and how the vehicle you will design is a reflection of those beliefs and values!

## Creating a Design Brief

The design brief is effectively a mission statement specifically for the intended design project. It's meant to succinctly clarify the objectives and goals of a program. With most design briefs, "less is more" and having some inherent flexibility and room for interpretation is good. However, it's important to be clear, concise, and have some key aspects well articulated (see Figure 3.2). The most important aspects are a definition of the target user, and an alignment with a broader mission statement or vision. For example, people who do not enjoy some aspects of driving are not Mazda's targeted users, and coming up with a design brief that doesn't prioritize the enjoyment of driving would not be aligned with Mazda's corporate vision or brand essence. A good design brief should also outline key goals with regards to budget, scheduling and timing, standards of the type of work to be done, and define an end deliverable. For example, a project that is intended for internal study in advance research may call for digital animation. This will involve a completely different timeline and budget than a project that calls for a one-to-one model intended for an engineering group.

Another essential component of a design brief is to identify the available technical resources versus required resources. In

△
**3.2** Essential components of a typical design brief
SOURCE: IMAGE CREDIT, JORDAN MEADOWS.

vehicle design programs, quite often there's a given engineering package layout based on technical assumptions for power plant, driveline, vehicle dimensions, storage, etc. A good design brief should also spell out a margin for these assumptions and how much deviation, evolution, and change can be considered appropriate for the project. For cost reasons, many corporations also use carryover components when assembling subsystems. So it's important to have an understanding upfront of what might be new versus what is pre-existing.

An effective design brief should also provide some quantitative standards for measuring the success of the program. For example, if there is a key competitor in a segment the vehicle needs to outperform from a technical standpoint, this should be clearly spelled out in the beginning as one of the objectives. The brief should offer a description of appearance objectives and aesthetic assumptions. This can be indicated often with some qualitative descriptors; for example, a Jeep should undoubtedly look tough, robust, and capable. Conversely, a design brief for Rolls-Royce would obviously outline a premium, upscale, and noble appearance as one of its objectives.

Vehicle design briefs often give quality metrics and targets to be considered, as these can have an effect on the aesthetics. Case in point, if a sports car had a weight requirement that would in turn affect the driving quality and performance characteristics, this would in turn affect the assumptions of materials. Perhaps

aluminum and carbon fiber would be required. This would have an impact on cost and govern the types of shapes that could be produced. If one were designing an interior of an open vehicle that needed to be waterproof, and offer extreme durability, this could also affect the look and feel of the design, the materials used, etc.

Finally, it is also quite helpful in some cases to have a brief that offers a table of do's and don'ts. This can be a simple document that provides a general list of what the project *is*, and what it *is not*.

Establishing a design brief is essential for getting started with a program, whether it is for an individual as a student or in any professional setting. In a sense it's a conceptual compass for the program. It gives both individuals and groups a clear reminder of where you are going and why. For large and small organizations, it represents a drumbeat or marching orders for a team. Freeform creative thinking is essential for the design process, but so are structure and discipline. A solid brief safeguards against delays, distraction, and miscommunication. The first two can easily plague individuals as well as large organizations. To that end, an effective brief can display the tangible application of goals and objectives for stakeholders. In academic settings, having an effective brief clarifies and focuses a student's learning objectives. It can ensure that you are working in accordance with a lesson plan. Having a well-written brief that you followed in a professional manner can showcase creative thinking as well as discipline for a potential employer. For independent designers and freelancers, a brief

enables clear communication between client and consultant. When combined with a contract, it can safeguard against cost overruns and awkward disputes with regards to payment and deliverables. For designers who are direct members of staff, a brief can allow for creative initiatives to be presented to the broader organization. It can be used as an effective tool for outlining and translating a creative vision to members of a development team who are not designers in a way they can appreciate and understand.

With all of these obvious benefits, having a good brief that a team or individual sticks to would seem to be common sense; however, the reality is it's easier said than done and many of the stumbling blocks and demons that pop up along the journey are a result of having a flawed brief or lack of discipline in following one. Pick any ugly vehicle that you hate to look at, and it's guaranteed that this is part of the reason for the eyesore! Conversely when a brief is well developed and followed faithfully, the results can be very effective.

# Mazda Motorsports
## and Furai Concept Mission and Brief

△
**3.3** Mazda Nagare concept, featuring body side textures inspired by nature
SOURCE: IMAGE CREDIT, MAZDA DESIGN.

In 2005, a relatively small but strong, spirited Mazda Corporation was about to enter a short lull in its image product cycles. Effectively there was a short gap between product launches for its brand Halo vehicles. At the same time, it identified a growing threat from emboldened challenger brands in Asia. To promote its sporty core message and further endear itself to enthusiasts, Mazda decided to embark on a series of **concept cars** to clarify its competency as a builder of lightweight, agile, fun-to-drive sports cars. The design team was tasked with creating a new **DNA** or visual language that

could be applied to the brand's forthcoming product. A decision by the design leadership was made to use movement in nature as a source of visual inspiration for the visual **aesthetics**. Vehicle designers have often sought to create a dynamic quality in their creations. In fact, there isn't much new about a vehicle looking like it's moving while standing still. However, fluid material movement in nature, how wind shapes sand in the desert, or how water can carve stone in a valley, were types of visual dynamic that had not been explored. The visual semantics of this phenomena

alluded to a sense of power and efficiency. At the same time, it communicated a sense of respect for ecology, making it socially relevant and aligned with the brand.

This design DNA was developed internally and titled Nagare, the Japanese word for flow, or the embodiment of motion. It was presented with the debut of the Nagare concept car shown at the Los Angeles Auto Show in 2006 (Figure 3.3).

The aesthetic language evolved through successive studies that were unveiled around the globe at international auto shows in the form of the Ryuga (see Figure 3.4), Hakaze (Figure 3.5), and Taiki (Figure 3.6) concept cars. Finally, the series culminated with the Furai concept that was unveiled at the North American International Auto Show in January 2008 (Figure 3.7).

I was fortunate enough to be a member of the design team that generated and executed this series of concepts. The design images which accompany this case study are a selection of my contributions to the Furai project. The creative process is a dialogue. In a studio environment many points of view need to be entertained and heard. The aesthetic language was very unique, especially when it took the form of literal textures stamped in sheet metal on the body sides of the vehicles. A criticism, however, was that the DNA would be even more relevant if it were supported by a technical function. This would effectively close a conceptual loop and unify the unique aspects of an aesthetic with its technical attributes. This would give the new DNA true substance, not just style but actual function (see Figures 3.8 and 3.9).

△△△
**3.4** Mazda Ryuga concept, sibling to the Mazda Nagare concept, that explored different textures and proportions
SOURCE: IMAGE CREDIT, MAZDA DESIGN.

△△
**3.5** Mazda Hakaze concept, exploring an all-road application of the design DNA
SOURCE: IMAGE CREDIT, MAZDA DESIGN.

△
**3.6** Mazda Taiki concept, using the DNA to express a sense of lightweight, flowing elegance
SOURCE: IMAGE CREDIT, MAZDA DESIGN.

△
**3.7** Mazda Furai concept, front tip up view
SOURCE: IMAGE CREDIT, MAZDA DESIGN.

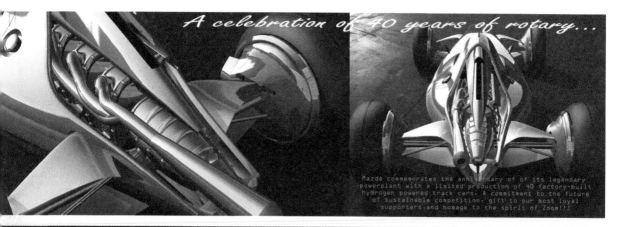

A celebration of 40 years of rotary...

Mazda commemorates the anniversary of of its legendary powerplant with a limited production of 40 factory-built hydrogen powered track cars. A commitment to the future of sustainable competition, gift to our most loyal supporters and homage to the spirit of Zoom!!!

j.meadows

△△
**3.8** Inspirational image created to focus discussion on rotary and racing heritage
SOURCE: IMAGE CREDIT, MAZDA DESIGN.

△
**3.9** Mazda Furai concept, front view depicting deep arrow channels sculpted to provide optimum flow while also communicating signature Mazda front-end graphic DNA appearance
SOURCE: IMAGE CREDIT, MAZDA DESIGN.

Mazda is a company of automotive enthusiasts. Mazda North America is also the home of Mazda's motorsports operation. On any given weekend in North America, more Mazdas participate in road racing than any other brand; this is evident in the spirit of every car that comes off the assembly line. Mazda racing is accessible to everyone from the weekend enthusiast to the professional driver. To put it bluntly, motorsports particularly built on endurance racing is deep in the brand's heritage. In fact, in the basement storage of the **R&D** facility there is a wonderful collection of vintage Mazda race-cars. These authentic examples of vehicles dedicated to winning provided excellent inspiration for the design team members but also offered the link for giving the new design DNA true substance (see Figures 3.10 and 3.11).

△△
**3.10** Mazda Furai concept, side view depicting dynamic body side sculpting and dominant rear wing
SOURCE: IMAGE CREDIT, MAZDA DESIGN.

△
**3.11** Mazda Furai concept, front three-quarter view showing the impact of different color on rear of body side sculpture
SOURCE: IMAGE CREDIT, MAZDA DESIGN.

Racecars are dependent on an understanding of **aerodynamics**. Physics define the shapes. At the same time the design industry is constantly moving toward digital and computer-aided models for testing and creating products. While aerodynamics need to be proven in a wind tunnel and ultimately on the track, the industry continues moving toward computational fluid analysis as a means of testing advanced aerodynamics. We found the shapes generated with this digital exercise to be progressive and new looking but also, not coincidentally, reminiscent of the

△△
**3.12** Mazda Furai concept, top view showing extreme arrow channels allowing optimal flow from front to rear and dramatic sculptural impact
SOURCE: IMAGE CREDIT, MAZDA DESIGN.

△
**3.13** Early Mazda Furai concept, study depicting progressive sculpture with modern forms complemented by colors inspired by classic endurance racing cars
SOURCE: IMAGE CREDIT, MAZDA DESIGN.

forms we were developing for the Nagare DNA. Computational fluid dynamics as used in the development of racecars became the link to unify the Nagare visual study with a high performance technical function (see Figures 3.12 and 3.13).

To further underscore the idea that the exercise would not just be a stylistic one, the brief for Furai was written to promote and proclaim a commitment to rotary technology. Due to cost and complexity constraints, many concept vehicles do not run. This is not essential if the focus is just

to communicate an evolving visual DNA. However, if world-class racing is part of your heritage as with Mazda, it's crucial to send a strong message about performance. To make good on this promise, the final concept was built on a Courage C 65 chassis. Going with a proven winner with one victory and nine podium finishes in 15 **ALMS** events during the 2005 and 2006 seasons meant the concept was predicated on real-world results (Figures 3.14–3.20).

△△
**3.14** Mazda Furai concept, presented at Mazda Raceway Laguna Seca, side view image captured as vehicle turns through hairpin corner
SOURCE: IMAGE CREDIT, MAZDA DESIGN.

△
**3.15** Mazda Furai concept, presented at Mazda Raceway Laguna Seca, rear three-quarter tip-up view captured on straight-way
SOURCE: IMAGE CREDIT, MAZDA DESIGN.

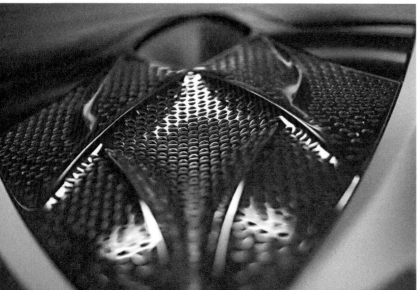

The team utilized an **ethanol-fueled** 450 hp RENESIS-based R20B three-rotor **rotary engine**. This ensured that from a technical standpoint the vehicle would honor the Mazda 787B, the first and only Japanese car to ever win the 24 hours of **Le Mans** in 1991 (Figures 3.21 and 3.22). It would also be a worthy centerpiece to the 40th anniversary celebration of the brand's use of rotary engines

dating back to the original Mazda Cosmo.

The team at Mazda generated a strong brief and followed it with extreme rigor. This allowed the Nagare series and Furai concept to be successful for three key reasons: (1) the brief was well aligned and timed with their business objectives. The brand needed to attract attention, and

communicate a commitment to rotary technology while working toward an eco-friendly future. Using visual inspiration derived from nature and combining it with an alternate fueled rotary engine allowed them to display a commitment to enthusiasts, while addressing critics who suggested high performance vehicles are environmentally irresponsible. (2) The very shape, **silhouette**, and positioning of the vehicle, halfway between high performance exotic and track-oriented racecar, were certain to communicate the brand message. Mazda still had the strong soul of a sports car and would continue to prioritize fun-to-drive vehicles that were competent on and off the racetrack (see Figure 3.23). (3) The visual DNA was not just a visual style, but was rooted in a technical function. That technical function was also aligned with the brand objectives. In fact, its heritage went back to the 787B.

◁

**3.20** Mazda Furai concept, interior featuring key controls integrated with removable steering wheel
SOURCE: IMAGE CREDIT, MAZDA DESIGN.

△△

**3.21** Le Mans-winning Mazda 787B 1
SOURCE: IMAGE CREDIT, MAZDA DESIGN.

△

**3.22** Le Mans-winning Mazda 787B 2
SOURCE: IMAGE CREDIT, MAZDA DESIGN.

△
**3.23** Mazda Furai concept, presented at Mazda Raceway
Laguna Seca, side view featuring Nagare body side textures,
accelerating through turn
SOURCE: IMAGE CREDIT, MAZDA DESIGN.

Looking forward, the group continues to promote its heritage with the LM 55. And even further in to the future with ultra-progressive design studies like the Kann electric racer (Figures 3.24 and 3.25).

When we view all of these cars together in a continuum, we can see the common thread of the brand DNA. In 1991, the organization needed to show credibility and proved it at Le Mans, one of the most legendary racing events. To this day, the 787B is a very special vehicle. Racecars and concept cars are very similar in that they both have the duty to promote

a brand message about their manufacturers and sponsors. The Mazda Furai was again a valuable communication tool that fulfilled its brief to spread the word about an evolving position for the brand. In 2014, the group reminded the world of its heritage and long-term message with another concept: the LM 55. And while this vehicle is obviously connected to the two examples previously mentioned, it was displayed and presented as a digital concept intended for the Sony PlayStation GT 6 videogame (Figures 3.26 and 3.27). This vehicle continues the tradition and introduces an entire new generation to the Mazda experience. With the LM

55, Mazda extends the range of its motorsports offerings from professional machines through to production based club racers, all the way to virtual competition cars for gamers. In all cases, talented designers following a strong mission statement and design brief provided the opportunity to enjoy the thrill of competition and soul the brand!

KAA

KAAN

△△
**3.24** Mazda continues to look forward considering how its racing DNA may evolve in the distant future with the Mazda Kann, experimental future electric racer, front view depicted
SOURCE: IMAGE CREDIT, MAZDA DESIGN.

△
**3.25** Mazda Kann experimental racer study took top honors winning the prestigious Designers Challenge at the Los Angeles International Auto Show in 2008
SOURCE: IMAGE CREDIT, MAZDA DESIGN

△△
**3.26** Mazda LM 55 design concept,
study executed for GT6 Sony PlayStation
videogame, rear view featuring advanced
spoiler and diffuser
SOURCE: IMAGE CREDIT, MAZDA DESIGN.

△
**3.27** Mazda LM 55 design concept,
study executed for GT6 Sony PlayStation
videogame, front view featuring evolved
Mazda DNA graphic
SOURCE: IMAGE CREDIT, MAZDA DESIGN.

IIIIIIIIIIIIIIIIIIIIIIIIIIIIIIIIIII

# Q&A

### ROBERT DAVIS

SENIOR VICE PRESIDENT, MAZDA
NORTH AMERICA AND HEAD OF
MAZDA MOTORSPORTS

*Robert T. Davis is Senior Vice President of U.S. Operations in Mazda North American Operations (MNAO) (Figure 3.28). Davis oversees all company operations within the USA that affect Mazda's customer and dealer-partner experience. Prior to this appointment, Davis served as Senior Vice President of Research and Development, responsible for vehicle research and development, manufacturing, product quality, and future product plans for all Mazda vehicles in North America.*

*Since joining MNAO in 1991, Davis has held various vice president and key management positions in sales, marketing, strategy, and field operations. Under Davis' direction, Mazda US Operations launched its sixth generation of models to strong customer and critical acclaim.*

*In addition to his business acumen, Davis is a true car enthusiast and racer. Under his leadership, Mazda became—and still is—the leader in North American motorsports, with the company participating in the sport at all levels from youth-karting to the highest levels of sports-car racing. He also oversaw the creation*

△
**3.28** MNAO VP Robert Davis
SOURCE: IMAGE CREDIT, MAZDA DESIGN.

*of the MAZDASPEED Driver Development ladder that was the first time a manufacturer had ever charted and supported young drivers as they climb the ladder from amateur to professional road-racing.*

*Prior to joining Mazda, Davis held multiple roles at Chrysler Motor Corporation, as well as spending time at a Chrysler dealership. Davis earned his bachelor's degree in Finance from Clemson University and an MBA from Georgia State University.*

**Question 1:** Who are your personal design heroes? Who do you most admire, and why? How does this key influence relate to the work you do for Mazda?

▸ As for personal design heroes, I could never answer with just one person as they do different things. When I look at design heroes, I'd have to say Harley Earl. He had not only a vision for design but also a feeling for the appropriate design for the appropriate market. He had a sense of balance and this is during the time when you could change the sheet metal on a car per model year almost. Inventive and ridiculously creative

△
**3.29** Mazda RX Vision concept, displayed at the Tokyo Motor Show in 2015, pays homage to classic rotary sports cars from Mazda's past while featuring the Kodo Design Language
SOURCE: IMAGE CREDIT, MAZDA DESIGN.

with jet inspirations, having responsibility for Buick and Oldsmobile and several brands. We all forget about the time when General Motors had all of these different divisions because a lot of them aren't around any more, but leading them, understanding them, and understanding the evolution of the design process and the business process is really interesting to me.

Then Vergil Exner out of the Chrysler camp is a good one. He did a lot of really interesting stuff that was the next generation of that era. I'm kind of a Chrysler guy family wise. [Having personal connections and previously working for the brand.] So we had them growing up and it was great

to see the design evolve all the way up through the mid-1960s and as they transitioned to the **muscle car** era, that was a side of Chrysler that was really interesting. On the modern design front, if you look at what Maeda-san has done particularly on the Asian and Japanese side of the business, it's really left an impression (see Figure 3.29).

When I look at the Tokyo Auto Show in introducing **Kodo**, he revealed this really awesome combination of not design for design sake, but also design that matches the brand presence and is not overdone (see Figure 3.30). And when I look at a lot of other Japanese design without naming brands it feels gimmicky

or overdone with a bit too much going on.

With Maeda, knowing his history and the history his father had with Mazda and what he did in the 1970s, there is a really nice connection. Even when you look at the Nagare era during your time with Mazda, Jordan, we went from a very fixed design style with prescribed guardrails almost, to Kodo where Maeda has allowed the designs to be very soft and very appropriate for each individual model. [Kodo is Mazda's current design DNA, preceded by Nagare.] So those are the three guys I would say would be my three role models or heroes on the design side.

△ **3.30** Mazda RX Vision concept, rear three-quarter featuring sophisticated emotional body side highlight composition
SOURCE: IMAGE CREDIT, MAZDA DESIGN.

**Question 2: While being a business leader you are also an avid racer; who are your personal motorsports heroes? Who do you most admire, and why? How does this influence relate to the work you do for Mazda?**

▶ As far as racers—there are a few that come to mind and some that I have had the pleasure of meeting. I think the combination of the personalities and what they've done on the racetrack is pretty inspirational. The first of that is Phil Hill. I had the chance to know Phil and just his pure versatility, and the fact that he existed during his era was amazing; just a pure role model not only in terms of how you race, but how you lived. The

next would be Mario Andretti. I think a man whose dream was the American dream. No money in the beginning, sometimes racing under an assumed name, being an immigrant and racing anything and everything he could get into. His versatility alone brought huge admiration from me because it just doesn't exist any more. I had the pleasure to meet him a few times. And then another is A.J. Foyt and Foyt was the opposite of Mario in terms of background but again was a very strong personality. I had a chance to meet him and his grandson when I lived in Texas. Both Andretti and Foyt now belong to this era of racers that had a different way about them. I remember reading that A.J. had a heart attack on his

bulldozer. And I thought, really? Bulldozing at your age? Wow!

From a pure racecraft standpoint and admiring the commitment, I grew up during the Senna era. So knowing what he did, and how he did it, and how much he was personally committed to his craft is really inspirational. And I think he was the first quintessential modern racecar driver. Fit, focused, just really doing everything very well. Certainly the numbers, the records, the results, and the way he went about things were very inspirational.

And then, finally, Katayama-san; He was a Mazda driver who recently passed away. I had the pleasure of meeting him a few times and he was always such a pleasant, really warm and nice man. And even though he was up in years, when he drove our cars recently, watching him on the track was so silky smooth. He was very Japanese in his driving style in the way that he was able to apply his craft. Even in a foreign country, with a foreign language, in a car that he had not driven in 20 years, it was a pretty impressive thing to watch. So when you see someone in that context, after not driving competitively but they still have a spark, is pretty inspirational.

Question 3: **Mazda has a history of creating great sports and competition cars. These vehicles in general have very clear and direct mission statements: "perform well and win!" How does the brand deliver a unique and differentiated experience to its customers from a design perspective?**

▸ Talking about great sports cars and competition cars that perform well, the easy thing for me to say is "Great road cars make great racecars!" If you look at all the driving principles of what composes a great Mazda road car, it's all about oneness: the *Jinba Ittai* experience. The connection between you and the car. [*Jinba Ittai* is a term relating to traditional Japanese horse-mounted archery.]

You and the car, the car and the road, or for that matter whatever visceral task that you take on. So I think that success in development of streetcars, whether it's the Cosmo sport 110 (see Figures 3.31 and 3.32), or the RX Three that I have now, to the first generation or all generation of RX Seven, even RX Eight, and

obviously Miatas, all tell you and show you why they are such great cars to drive on the street and why they'll bring you joy on the street.

And what drives people to make them race cars? You know we as a brand have 53 percent market share in amateur racing. So not discounting the hard work of a lot of people to get that done, but the base car is so good that it enables that 53 percent to capitalize on

△△
**3.31** Mazda Cosmo 110, front view, illustrating roots of Mazda's sporting DNA
SOURCE: IMAGE CREDIT, MAZDA DESIGN.

△
**3.32** Mazda Cosmo 110, rear view featuring lightweight, poised athletic, sporting stance common in almost all sports cars to this day
SOURCE: IMAGE CREDIT, MAZDA DESIGN.

△
**3.33** Mazda Miatas racing through turn one at Mazda Raceway Laguna Seca, offering thrills and excitement for spectators as well as drivers
SOURCE: IMAGE CREDIT, MAZDA DESIGN.

it. And then the position that we're at in the marketplace that allows those cars to be available to enthusiasts. Sure, a lot of people would love to be driving a Porsche or late-model BMW on a racetrack, but the price and affordability to get there are an issue. So the combination of the cost, the car accessibility, and the great programs the motor sports guys developed has driven success. But if you boil it all down to the very first step, it's to develop a great car. And this requires a clean, clear mission statement. A concise vision of where you want to go. This allows everyone

to execute the goal and deliver it. And I think the latest iteration of the Miata is a great example of that (Figure 3.33).

**Question 4: When designing vehicles other than sports cars, what type of customer needs, aspirations, and emotional experiences does the design group consider in delivering the Mazda experience?**

▶ As far as vehicles other than sports cars, I think what consumers need is to satisfy their aspirations. Yes, it's true that not everyone can own a sports car in every stage of life,

but the influence it has on your life, and what it makes you feel like are a very important thing. That puts a smile on your face. So, for example, I spend almost two hours a day to commute in my car. And I've got a really nice car at the moment, a Mazda Six GT. Now my driveway at the end of my commute is really fun, and the drive at the beginning part is really fun. And driving that car on an exciting twisty road is really enjoyable and puts a smile on your face. For guys like us, when you have that type of situation, you can focus on the things you want to change. For example, my

△
**3.34** Four iterations of Miata and MX 5 cup racers, making the Mazda experience available to the masses through successive generations
SOURCE: IMAGE CREDIT, MAZDA DESIGN.

new pet peeve is when you hit the sport button (to enjoy the best part of my commute) that lights up yellow. Now it really ought to light up green! Sport mode should be a positive thing, not cautionary. Just understanding all the small details in a car and how they add up to a great experience can make a big difference at the end of the day.

I think that also you see the heritage from the racing side come through in the details of some of the road cars, whether it's in the lighting signatures and touch

points. Ken and Julian (in design leadership Mazda North American Operations) and our design team use the 787 as inspiration along with a lot of our other racecars. Julian Montouse's specialty is interior design. And I've known him to sit down and look at the interior of racecars because the layout is very functional. It's all very dedicated to being lightweight and designed around being able to find things with your hands with a helmet and gloves on at 2 o'clock in the morning at Daytona or Le Mans racing down the straights at

200 miles an hour. So those are things we all see.

I think from a student of design standpoint, or for the readers of this book, it's about understanding whatever the passion is and letting that guide the development of ideas and execution. Whether your passion is motor sports, or horses, or whatever, just allow yourself to go with it and be taken away. Because I think there is beauty in all the natural and functional elements coming from motor sports, it's just about transitioning that beauty into what your style is. For example, motorsports evolving in the aerodynamic age from 787B, to Furai, and even in the PlayStation LM 55 car, that beauty of airflow is very interesting. Now there are people who execute it the right way, and there are people who execute it the wrong way. For example, some of the WEC (World Endurance Challenge) cars are absolutely stunning and beautiful. And then some of them are not. Obviously they're doing very similar things in the wind tunnel, right? But why do some look good and others don't? As Weldon Munsey [Manager for Mazda North America Partner Affairs] once said: "Everyone has to put plastic in a car but that doesn't mean you have to use cheap-looking plastic!"

There are obviously people who want to have beauty and function. So someone has to commit to it as a priority. And so you come to the same conclusion: WEC, like Formula One these days, they all have very strict parameters that they are working within. And as in **F1**, some of the cars are absolutely hideous to look at, and others really look great. In both cases someone is making a decision, and the car that looks good, and is actually competitive, that's a great combination.

**Question 5: Looking toward the future while considering zero emissions and autonomous technology, what are the biggest challenges facing designers in delivering high performance vehicles for racers and enthusiasts? What is your advice to designers to conquer those challenges?**
▶ I think what students and readers of this book have got to consider is how to keep the

passion in products. If we allow ourselves to listen to Wall Street, or trade publications, or people with different agendas, we will wind up designing the best compartment for a La-Z-Boy or Barcalounger. And I don't think any of us want that to happen! And while the population, and the mobility space, are going to change significantly in the coming years, there is still plenty of room for excitement. There will still be plenty of room for sports; plenty room for the passion to execute it.

So I think it's more around staying clear to your vision, trusting instinct, and staying clear with what you're going to do. On the **autonomous** side, our position is how can we use the technology to aid the driver? To make his or her drive easier or more comfortable, not taking the responsibility of the journey completely away from them.

*We're not about just moving from point A to point B.*

Now if that's lane assist, or crash avoidance, or whatever the technology involved, we still want it to be fun, and we still think there will be plenty of room in the marketplace for that. Now some companies will choose to go another route. But I think of that as an alternative to a subway or a taxi, but not a Miata (Figure 3.34)!

Notes
1  www.daimler.com/company/strategy/. Web (accessed March 2016).
2  www.mazda.com/en/about/vision/. Web (accessed March 2016).

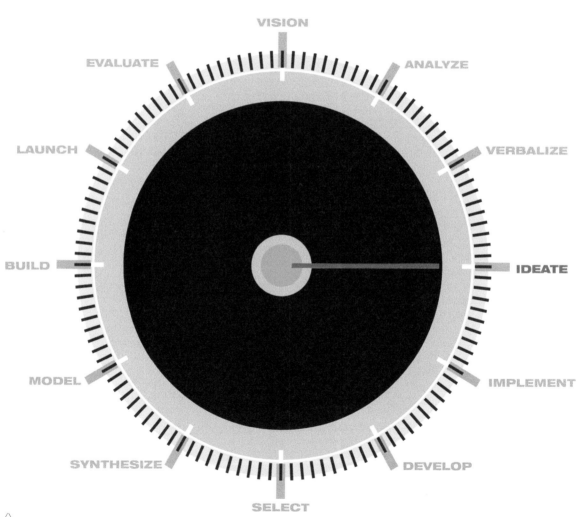

VISION

EVALUATE     ANALYZE

LAUNCH     VERBALIZE

BUILD     **IDEATE**

MODEL     IMPLEMENT

SYNTHESIZE     DEVELOP

SELECT

△
**4.0** Process locator gauge
SOURCE: IMAGE CREDIT, JORDAN MEADOWS.

**IDEATE**
CREATING A UNIQUE VISUAL DNA

**CHAPTER 4**

# IDEATE

## Creating a Unique Visual DNA Based on a Design Brief

### Get Going! Staring at a Blank Page

This section of the book is dedicated to the vehicle ideation process. This is the exciting and sometimes daunting point in the journey when we actually put pen to paper and commit to creating a concept. Many authors, screenwriters, and novelists speak in great depth about the sheer horror of staring at a vast white page and having to generate brilliance. For many designers, this stage of the game can be just as terrifying. Fortunately though, if we've followed some of the methodologies in this book, there's nothing to fear. We've initiated the effort with a well-defined vision. We've given ourselves a well-tuned compass for the concept in the form of a **design brief**, and a **mission statement**. And as we discovered in Chapter 2, we have armed ourselves with a detailed understanding of our targeted users' wants and needs. So moving forward should be easy!

Good design is about satisfying not only the head but the heart as well. As designers we've got to appeal to functional concerns as well satisfying deep-seated psychological desires. Once we've developed a rock-solid narrative that matches a defined archetype with the corresponding wants and needs, we can then get on with the business of designing that solution which provides the user with the vehicle of their dreams. And since we're now about to get tangible, the next step is to identify a literal architecture and structure for the idea.

The modules in this book are very similar to most product development strategies. Each of the ideas generated is intended to interlock and align with each step along the way, giving meaning to the next (see Figure 4.0). So before we actually start creating shapes, lines, and beautiful curves and sections, we need to develop an architecture and package for a vehicle, and that architecture should be defined and aligned with all of the work we've done up until now. The aim at this stage of the game is literally to let the archetype of the user define the architecture of the vehicle.

## Explore Various Vehicle Packages and Technical Solutions Based on the Needs of Your Target Customer and Market Opportunity

If a particular **vehicle segment** or product type has not been given or defined in the initial design brief, then the first stage is to establish an understanding of the scale and size of product that your user would want or need. Vehicle markets around the world define vehicle classes in several different ways. For example, in Europe, there is a letter designation that aligns the vehicle weight, width and length to an [A], being a segment that is subcompact. The [B] segment is around 4 m long. The [C] segment is around 4½ m long. The [D] segment is around 5 m long. Mercedes' Smart being an example of an [A] segment car. The Ford Fiesta being an example of a [B] segment car. The VW Golf is an example of a [C] segment car, while the BMW 7 Series is an example of a [D] segment car. Different regions have different segments to consider. In North America, midsize trucks versus full-size trucks are segmented, the Toyota Tacoma being an example of a midsize versus the Ford F150 being an example of a full-size. Many vehicles other than cars are classified in similar ways. For example, the cruiser class of motorcycle being typified by Harley versus the sport bike class being typified by the Ducati superbike. Furthermore, these motorcycle classes can be broken down into subcategories based on their power, weight, and size. Regardless of the type of segment, the key is to focus your energy moving forward with an understanding of a relative scale, size, and basic purpose the project requires. With a vehicle segment and product type in mind, then the focus of the work can be shifted to what type of architecture should be implemented.

The dictionary defines architecture as such:[1]

1. the art or science of building; specifically: the art or practice of designing and building structures and especially habitable ones.
2. a: formation or construction resulting from or as if from a conscious act <the architecture of the garden> b: a unifying or coherent form or structure <the novel lacks architecture>.
3. architectural product or work.
4. a method or style of building.

5. the manner in which the components of a computer or computer system are organized and integrated.

All of these definitions apply to vehicle design. However, the common term used to describe the vehicle's architecture is known as "packaging." This simply refers to how the designers have created a means to tackle three main issues: (1) people; (2) the stuff they need to carry and use; and (3) the power for propelling and controlling the driving experience. No matter how daunting and complex vehicle engineering can seem, it all boils down to these three simple subjects: people, power, and stuff. Within this,

△
**4.1** The key elements of vehicle packaging
SOURCE: IMAGE CREDIT, JORDAN MEADOWS.

however, **vehicle engineering** can then be broken down into five corresponding areas of activity. Arranging and orchestrating these systems is both a science and an art that even the most gifted can take an entire career to master (see Figure 4.1).

These basic systems include first and foremost how the driver and the occupants are literally and figuratively placed in the equation. Second, the power and energy systems used to motivate the vehicle. Third, the basic features and function of the interior used to organize the user experience. The structure and body that tie the entire equation together are the fourth. And fifth are the wheels, tires, and suspension elements delivering a specific ride type. There are several publications that cover **vehicle architecture** and engineering in-depth. One of the best reference books for designers is entitled: *H Point: The Fundamentals of Car Design and Packaging.* Co-authored by Stuart Macey and Geoff Wardle, it provides a wealth of information on the subject. Packaging, however, is only one component of the ideation process that we will cover in broad overview in this chapter of this book.[2]

With 100+ years of automotive and vehicle engineering development there are some remarkably good packages that now exist. And again, if your design brief has already prescribed one, embrace it. Analyze the things that it does well, or poorly, and consider the opportunities for how it might be improved or evolved. These enhancements or updates to the package should ultimately be aligned with the story or

**narrative**. For example, for the longest time, the Jeep Wrangler in two-door form was the gold standard for off-road vehicles. However, when Jeep identified a want or desire for a particular type of user to carry more passengers with them, they investigated the idea of updating the traditional Wrangler two-door package with a longer wheelbase, seating for four, and two additional doors. The result was the Wrangler Unlimited, which now outsells the traditional two-door Jeep by a significant margin. As is the case with many traditional vehicle packages, the fundamentals can be used as a great point of departure. In professional scenarios, vehicle development teams spend a good amount of time analyzing and evolving their packages, and in many cases small incremental changes can lead to a radically different user experience.

With this in mind, sketching out the package with its main architectural volumes is one of the first steps in familiarizing yourself with what goes where, and where the opportunities for improvements might be. If the package is pre-existing, then iterative sketching with transfer paper and overlays can quickly establish some of the basics. And then if adjustments or package enhancements need to be made based on a specific user want or need, one can quickly see in side view and plan view what the issues may be. For example, if you were designing a sports car for an enthusiast, and realized there was a want or need for the vehicle to be more efficient than its predecessor or competitors, one might propose that the occupant package be lowered to gain a

more aerodynamic **silhouette**. Shifting the driver **H-point** [hip point] downward may be a proposed solution. Conversely, perhaps you wanted to propose a sports car for a user who had a want or need to carry gear, sporting equipment, or luggage. Sketching a side view silhouette that showed the occupant's H-point raising slightly along with the wheelbase extended slightly may be a way to increase the vehicle's functionality by providing more storage room. In both of these cases, the amendments to a traditional sports car package will be based on specific wants and needs established in the design brief and the narrative.

Quickly sketching out these options in side view, plan, and end view establishes the driver's and occupant's location, locking in the most crucial component of the vehicle architecture. Philosophically, and literally, the user is the center of the endeavor. So it's crucial at this point to see mannequins placed in the vehicle having a posture that is aligned with their wants and needs. If you're intending to do a vehicle for a Hero **archetype** who is interested in a high performance vehicle, then the occupant should be drawn in the shape of a low-slung sports car with an H-point quite low to the ground. Conversely, if you are doing a vehicle for an Explorer archetype who is interested in off-roading, the occupant should be drawn in an upright posture that affords command of the road with great vision angles for tackling unique and unknown off-road driving scenarios. The vehicle architecture stems from the posture and location of the user in both cases and establishes a set of assumptions for all of the remaining subsystems. Vehicles are driven and ridden, but they are also worn. The architecture of the vehicle's body is predetermined by the user's posture, placement, and priority (see Figure 4.2).

## Consider the Powertrain Implications on Basic Packaging. It's Physically Motivating the Vehicle!

Once you have figured out what type of position is best for the driver and occupants, the second step is to select **powertrain** and location if that has not already been predefined in your package. Many conventional packages, even in hybrid application, call for some sort of combustion engine generally located in the front of the vehicle; this is accompanied by a transmission that drives the wheels. In most vehicles, two-wheel drive is utilized. But in several applications all-wheel drive is utilized. With combustion engines, powered conventionally by gasoline or some alternative fuel, placing the main block longitudinally along the length of the vehicle, or mounting it transversely along the width of the vehicle can have a great impact on the vehicle's architecture. And as you guessed, if the powertrain selection, location, and mounting have not already been defined or given to you, it's best to establish where they are packaged based on your user's wants and needs. For example, if your Hero archetype user is obsessed with performance, then there's no better place to put that engine and transmission than just behind his back in a very low-slung, mid-engine, high performance package. After all, the weight distribution of a mid-engine set-up is very well

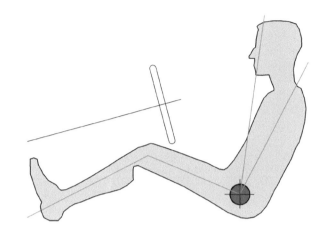

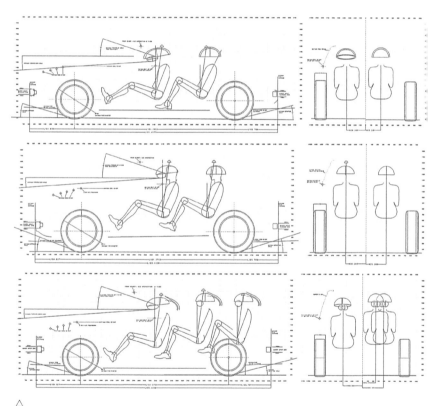

△
**4.2** Side view of typical driver mannequin with H-point indicated by the circle. Lower parts of image illustrating different H-point locations and driver and occupant configurations for a typical sedan seen at the top, a typical SUV depicted in the middle, and a common people mover package, featuring a third row of occupants in the lower package side view drawing
SOURCE: IMAGE CREDIT, JORDAN MEADOWS.

balanced, concentrating the mass between the wheels and offering the best handling. Conversely, if you are designing for an Explorer archetype who needs a high degree of vehicle capability, then a front longitudinally mounted engine driving all four wheels may be the way to go.

**Battery electric** powertrain technologies offer vehicle architects a good degree of freedom for packaging. By containing the batteries along the floor of the vehicle and mounting the motors within the tire envelopes at the ends, a good degree of flexibility for various occupant placements can be achieved. In some cases, the batteries can be mounted down the center of the vehicle utilizing a backbone-style chassis that allows the driver's H-point to be placed even lower. In any case, selecting the powertrain technology and coming up with an optimal layout should be the result of understanding the vehicle usage scenarios. For example, if you were designing a long-range traveler, and the user required the flexibility of not being limited by battery range, some sort of hybrid configuration might be best. Or alternatively, if that same user also wanted a **zero emissions** solution, then hydrogen could be explored. In that case your vehicle packaging should also take into account the special needs for packaging hydrogen fuel tanks safely.

Optimally placing the vehicle's occupants in position, and selecting an appropriate powertrain will already give you an impression of the vehicle's silhouette. As you continue sketching the overlays, consider

the **occupants' sightlines** and vision angles. In almost all cases having a windshield touchdown and angle can be a big part of the user experience. The faster windshield is more **aerodynamic** and appropriate to sports car configurations, while a more upright windshield placed closer to the driver is more appropriate for high H-point and all-road vehicles. Also, be mindful to sketch the vehicle architecture in end view and plan view. Having a wide vehicle with three across seating and a very upright body section will give a drastically different impression than having two occupants placed closely together.

Finally, in creating these vehicle silhouettes that can already send very strong messages about the intent of the usage, it's time to consider where the rubber meets the road. And as with all of the other key elements of the vehicle package, if it hasn't already been predefined and given to you in the design brief, then make sure to align the wheel and tire and suspension capability to fit the customer narrative.

Remember our Explorer archetype who demanded high capability? It stands to reason that this user would need a very tall tire and suspension set-up with a good degree of travel. This would ensure that that vehicle can take on multiple situations. Conversely, for the Hero archetype who was obsessed with high performance on the track, it goes without saying that their package would require a very low-profile tire with a wide aspect ratio, providing good grip in a fairly wide track in end view.

## Structure and a Framework for Vehicle Architecture

Through sketching various overlays, you are naturally coming up with a silhouette that is defined by the user's wants and needs and indicative of a particular body structure. Most vehicle body structures are created in two ways: the original being a frame structure that carries the powertrain, occupants, and wheel and tire and suspension systems. This frame is created with all of the rigidity and strength necessary to effectively provide basic mobility. The glass, sheet metal, bodywork, and interior of the vehicle are placed on top of the frame. The industry term for this type of construction is called **body on frame**. It's used most frequently on trucks and SUVs. It also offers a great deal of flexibility since most of the engineering-intensive components can be grouped together and carried with the frame, while the bodywork can adapt and change to different styles and applications.

A second and more commonly used structure by the industry is known as a **uni-body construction**. This application is used for passenger cars, light SUVs, and some light duty vehicles. In this case some of the actual bodywork is loadbearing while only the aperture elements such as the doors, hood, and deck lid are nonstructural. And third, some high performance vehicles use a cage type or **monocoque** construction. These structures can be fabricated in high tech carbon fiber, or welded and bonded aluminum. This allows for a good degree of rigidity while also being lightweight. With this application all of the bodywork can also be made as light as possible. While continuing to sketch more overlays, consider that the body construction will also have an implication on the look, shape, and feel of the vehicle. "Body on frame" cars tend to be taller and require a higher H-point, whereas the low-slung sports car may require a lightweight monocoque construction.

As the ideation process continues, consider how your occupant gets into, and sees out of, the vehicle. It's important to consider closures, apertures, and occupant vision early on as this will inform the graphics and shapes of the windows and daylight openings.

Also, begin to define the cargo space for the vehicle. Again, this is also based on the usage scenarios and customer narrative if not indicated in your initial design brief. The user's ability to carry cargo will also have a massive impact on the silhouette, and look of the basic packaging. If your intended user needs to go to Home Depot and carry a 4 x 8 sheet of plywood or several other things to care for their home, then having provisions accommodating these objects could be essential. For example, a large open bed on a pickup of a certain width and length. Or a rear hatch of a certain width, length, and height coupled with folding rear seats, for example.

Whether fast or slow, vehicles are moving objects that are influenced by aerodynamics. The optimal speed of the vehicle and impact of aerodynamics on its shape are also a result of the design brief or user narrative. If the low-slung high performance sports car needs to compete, it stands to reason that it may also have spoilers, diffusers, or wings. Conversely, perhaps you're designing a vehicle for low-speed city usage where aerodynamics is less of a concern. In this case having an optimal aerodynamic shape would be less of a necessity.

And, finally, as you are continuing rough overlays to establish packaging and vehicle architecture, you must also consider safety requirements. Some of these can be defined by laws. Others can be defined by corporate objectives for insurance compliance. In any case, if it's not prescribed in the design brief, one must ultimately take into account safe operation for the vehicle.

## PACKAGE

LENGTH    5130
HEIGHT    1490
WIDTH     1990

WHEEL BASE   3187

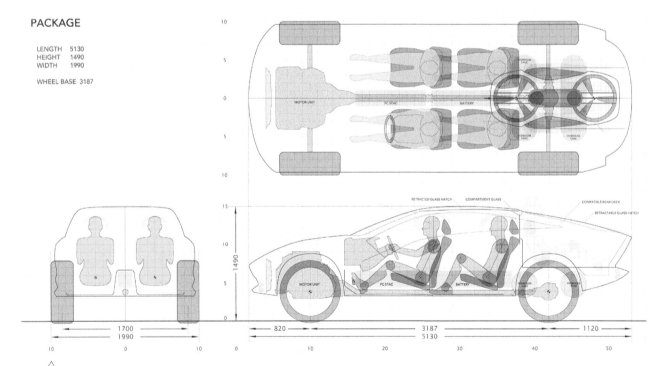

△
**4.3** Package drawing featuring occupant and tactical layout for Toyota Cross Cruiser concept. Note provisions for special hydrogen tanks and battery packaging required for hybrid hydrogen powertrain.
SOURCE: IMAGE CREDIT, KEN NAGASAKA.

△
**4.4** Supplemental copy of vehicle inspirations that depict the desired Toyota Cross Cruiser concept, user experience
SOURCE: IMAGE CREDIT, KEN NAGASAKA.

## MARKET PREDICTION

Although traditional internal combustion engine vehicles will be still dominant in 2030, the market structure will be more diversified since new power train vehicles will increase in the market. This complex age would be the selection period of next power train.

2015    2030    2060

120

110

100

90

80

Million Units

FCV

EV

HYBRID

GAS

?

DIVERSITY

World Car Sales

**HOW CAN FCV APPEAL**

The market is going to be more complex and competitive, since prices will be closer and each cars will spread to the market. Therefore FCVs need to be distinguished from competitors with strong massage to consumers who are confused by complex lineups.

▶ IT SHOULD NOT BE JUST A ECO CAR.

△
**4.5** Image illustrating hypothesis of increased demand for zero emissions vehicles of how to add emotional appeal to the genre
SOURCE: IMAGE CREDIT, KEN NAGASAKA.

Having front and rear crush zones and bumper beams, for example, will also impact the design and shape of the bodywork. Many modern cars also have airbags and significant structures to protect the occupants in the event of a rollover accident. These elements can affect the shape of the daylight openings, sections of the roof and rails, and thickness of the pillars. It's important to note these regulatory and safety elements early in the process. Taking them into account and designing a unique way to deal with them can separate your vehicle from its competition (see Figure 4.3). The package shown in Figure 4.3 is how Ken Nagasaka developed vehicle architecture to support his project goals and objectives.

In summary, having a flawed or inappropriate package is the easiest way to disappoint a user. When the architecture and packaging are arranged well, the vehicle will have a great basis to conduct further design work. Getting the basic compositional elements in place is the first step in creating tangible solutions, and crucial to the success of the vehicle!

Having defined the vehicle package and established a framework for the functional objectives of the vehicle, the next step is to identify and collect imagery that suits the user's tastes. This visual inspiration should also be aligned with the user's **storyboard** outcomes and aspirations. And even though at that stage of the process, the aim was to provide an emotional experience, a designer should assemble images that express those ideals in the form of objects. These images will provide inspiration and reference for the vehicle's specific morphology (see Figures 4.4 and 4.5).

For example, if one is designing a vehicle for a family led by a female, mid-30s Gen X (Caregiver archetype) with two children, a vehicle package with third row seating for six may be desirable. The question then becomes what type of imagery would be appropriate for that project? What types of shapes and forms would resonate with that user? Again, the narrative provides the means to balance and align the values of the brand, and the buyer.

In Chapter 2 we learned that the world we create around us can be viewed as an extension of our psyche and persona. We could then theorize that all the choices that our Caregiver archetype user would make and the nonautomotive objects that she would enjoy, support her narrative of being a person who strongly aspires toward looking after her group. If you were to select a piece of furniture for her, you'd be sure that it would be warm and inviting to a group of people. The shapes and edges would likely be round and soft to accommodate children. The same would go for the other objects in her life. With these assumptions, we can assemble a collection of images and things this person would enjoy. And provided these images were aligned with the values of the brand, it would make for a great image board.

Having established the basic architecture and armed with the

relevant imagery, we're ready to explore proportion, form, graphics, and lines. Modern-day vehicles are complex mechanical organisms. The frame and structure are the bones. The powertrain and key elements are the heart and organs. With these in place, the next stage in their creation is to define the muscle, flesh, and skin. Any aspiring artist who's taken a figure drawing class knows that one of the basic techniques in capturing a subject visually is to imagine that the skeletal structure is supporting its mass. The process of sketching vehicles is exactly the same. As you put pen to paper, always consider two main things initially. First, that the organism that you're sketching has a structure, or architecture, and, second, the shapes, forms, and graphics that you visually convey should send a message that is aligned with the narrative and corresponding imagery.

### Explore Unique Visual DNA for a Vehicle Based on Objectives Established in Your Design Brief

One technique for getting many ideas out on a page for evaluation is a process of **rapid visualization**. This involves very quick sketching to block in the basic compositional elements of the vehicle. In this regard, it's a challenge to see if you can convey an attractive message in as few lines as possible. Provide only the absolute minimum amount of information to convey the theme. Generate a wide variety of memorable silhouettes that align with the given visual inspiration and user narrative (see Figures 4.6 and 4.7). They can have the simplicity of woodblock prints or can be conveyed in simple black-and-white images with the immediacy of signage. Here again

the figure drawing methodology of quickly capturing an impression or gesture is a great analogy.

In the mid-1970s perceptual psychologist James J. Gibson developed a widely accepted theory that some objects by their very nature exhibit the possibility of an action.[3] "Affordances," he theorized, describe the relationship between an object and stimuli; for example, a wheel naturally affords turning. This theory has evolved into different sub-definitions that are applicable to design, with the key being that an object or shape can communicate naturally its possible action. Don Norman in his book *The Design of Everyday Things* further developed the theory to take into account that a user can interpret these affordances differently based on their own point of view and context.[4] **Semiotics**, the general study of signs and their meaning, should also be considered when visually communicating an intended action or behavior. For example, a red tail lamp on a vehicle is a commonly recognizable and understood symbol. While cranking out many sketches, take into account the vehicle narrative and consider the inherent messages that the basic shapes may afford. Use the commonly understood symbols to convey your message. Then challenge them when necessary. A cube, for example, will in most cases look stable. If

▷

**4.6** Toyota Cross Cruiser concept, inspirational loose rapid ideation sketches capturing key themes, early exploratory work done to capture unique visual DNA for the vehicle based on objectives established in a design brief
SOURCE: IMAGE CREDIT, KEN NAGASAKA.

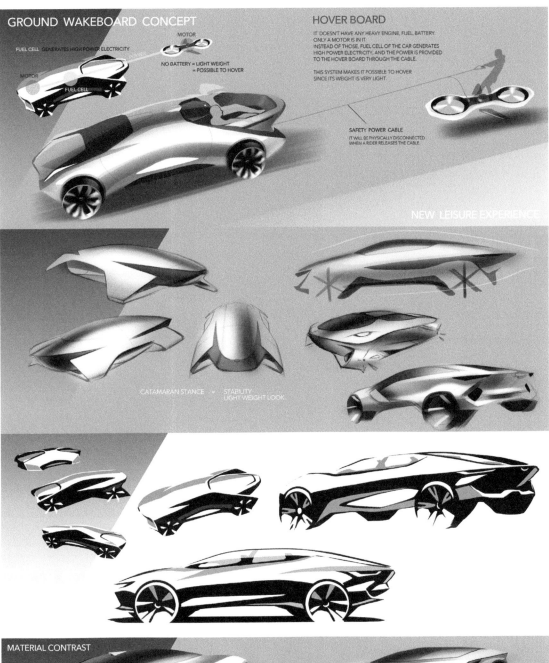

# GROUND WAKEBOARD CONCEPT

FUEL CELL GENERATES HIGH POWER ELECTRICITY

MOTOR

FUEL CELL

MOTOR

NO BATTERY = LIGHT WEIGHT
= POSSIBLE TO HOVER

## HOVER BOARD

IT DOESN'T HAVE ANY HEAVY ENGINE, FUEL, BATTERY.
ONLY A MOTOR IS IN IT.
INSTEAD OF THOSE, FUEL CELL OF THE CAR GENERATES
HIGH POWER ELECTRICITY, AND THE POWER IS PROVIDED
TO THE HOVER BOARD THROUGH THE CABLE.

THIS SYSTEM MAKES IT POSSIBLE TO HOVER
SINCE ITS WEIGHT IS VERY LIGHT.

SAFETY POWER CABLE
IT WILL BE PHYSICALLY DISCONNECTED
WHEN A RIDER RELEASES THE CABLE.

NEW LEISURE EXPERIENCE

CATAMARAN STANCE → STABILITY
LIGHT WEIGHT LOOK.

## MATERIAL CONTRAST

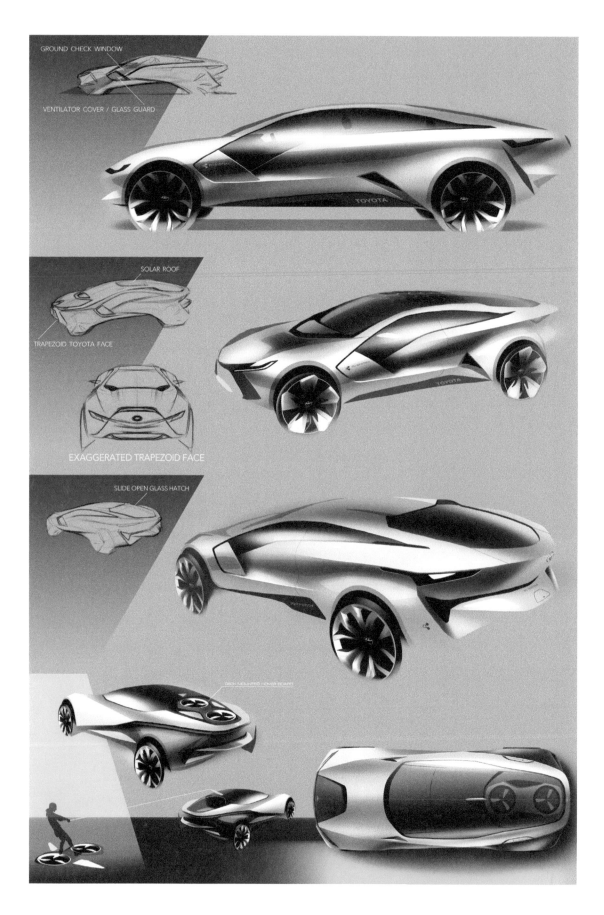

GROUND CHECK WINDOW

VENTILATOR COVER / GLASS GUARD

TOYOTA

SOLAR ROOF

TRAPEZOID TOYOTA FACE

EXAGGERATED TRAPEZOID FACE

TOYOTA

SLIDE OPEN GLASS HATCH

DECK MOUNTED HOVER BOARD

stability is a value that is aligned with your core message, then a cubic compositional element is the perfect way to go. If a soft, friendly character is a value that you're trying to convey, sketching graphics, shapes, and forms with a good degree of roundness may be appropriate.

Assembling these various shapes around your vehicle architecture in a memorable way that is aligned with your narrative is the designer's main aim at this point. Most of the architectural information is conveyed in side view. Cars are very easily understood in side view. For this reason, it is quite helpful to rattle off various iterations in side view (Figure 4.8). However, much of the personality of a vehicle is understood in front view. Again, we refer to the fact that from a psychological standpoint, our natural tendency is to anthropomorphize the object. So in conjunction, it is important to spend a good amount of time assembling the various compositional elements in front so they communicate your intended message. Front views convey a brand message that can also be important.

Front views also have a high degree of technical and regulatory features that can define the composition (see Figures 4.9 and 4.10). Added to this, the percentage of those elements relative to the overall scale of the

◁
**4.7** Toyota Cross Cruiser concept, intermediate theme development exploring proportion, form, graphic and signature lines. Note the ever important face of the vehicle is being investigated, with unique selling points such as the side window shape that allows for increased visibility
SOURCE: IMAGE CREDIT, KEN NAGASAKA.

△ △ △
**4.8** Toyota Cross Cruiser concept, final two-dimensional side view elevation drawing done in conjunction with the package in preparation for three-dimensional validation via model-making
SOURCE: IMAGE CREDIT, KEN NAGASAKA.

△ △
**4.9** Toyota Cross Cruiser concept, front three-quarter sketch, showing design and appearance intent for the front end graphics and overall stance of the vehicle
SOURCE: IMAGE CREDIT, KEN NAGASAKA.

△
**4.10** Toyota Cross Cruiser concept, front three-quarter rendering of digital model
SOURCE: IMAGE CREDIT, KEN NAGASAKA.

△
**4.11** Toyota Cross Cruiser concept, rear three-quarter theme rendering depicting subtle surface articulation and graphic details
SOURCE: IMAGE CREDIT, KEN NAGASAKA.

▷▷▷
**4.12** Toyota Cross Cruiser concept, three-quarter rendering of the digital model
SOURCE: IMAGE CREDIT, KEN NAGASAKA.

▷▷
**4.13** Toyota Cross Cruiser concept, side view rendering of digital model
SOURCE: IMAGE CREDIT, KEN NAGASAKA.

▷
**4.14** Toyota Cross Cruiser concept, rendering of digital model featuring overboard concept and design provision for storage in the rear compartment
SOURCE: IMAGE CREDIT, KEN NAGASAKA.

vehicle in that view can be quite significant. For instance, the average midsize sedan will require a front license plate, lighting of a given size, air intakes, bumper beams, and safety features, on a relatively compact area of the vehicle. Added to this, there is a very clear intent to make a brand statement. The designer must arrange all of these elements in such a way that a particular character is conveyed. And because of this, a good degree of personal taste is required to attract potential drivers. Needless to say, front ends are an area of the car that demand a good amount of attention.

Conversely, the rear end of the vehicle can leave you with an impression that is lasting and memorable (see Figure 4.11). The tail lamp design is another opportunity to send a key message. Quite often the rear of the vehicle will need to fulfill defined requirements from an aerodynamic perspective. Accessibility to cargo must also be considered and rear license

plate placement and brand signature are essential. And again with some applications such as a sedan, all of this is going on a relatively compact area of the vehicle. During the vehicle ideation process it is helpful to sketch out these various elements in different arrangements to explore multiple effects.

While the front view, rear end view and plan side view are coming together, one can also explore perspective and quartering views that can convey a dynamic quality and gesture of the vehicle. Perspective views are also effective in roughly outlining key features, and graphics. Quite often these are also very helpful in understanding how the sections and formal nature of the corners of the object can be executed, and how the vehicle sits on its wheels. A designer should assemble the key isometric views and constantly reference them so that all are aligned and derived from the same idea (see Figures 4.12–4.16). Ultimately these views should culminate in

the form of a concise side view elevation package drawing. In the classical automotive design studio environment, these package drawings were executed with tape in **scale model** form initially and then in full size. Using tape gives a graphic quality that can be clearly understood and recorded with digitizers. And while quite often they are today

**4.15** Toyota Cross Cruiser concept, final image of user enjoying experience of Cross Cruiser and overboard concept
SOURCE: IMAGE CREDIT, KEN NAGASAKA.

performed with digital media using two-dimensional computer-aided tools, tape drawings are still a fantastic way to communicate the key fundamentals of a vehicle's design. It is the key graphic representation used to communicate a design's relationship to its architecture and construction. It is also the main means for translating the two-dimensional representation of a design into **3-D**, be it virtual data or **clay modeling** form.

## Begin to Explore Surface Language

Once the basic elements of the vehicle's composition have been established, and the designer is confident that the proportion and execution of the volumes over the vehicle's structure feel aligned with the goals and objectives of the project, the next step is to consider the development of the **surface language**. In this regard the transitions, nuances, and subtleties of the vehicle's sculpture can have a profound impact on its impression. Consider the different sections and the different messages they convey. Again, we refer back to the earlier described theory that the surface language can also convey a message. Having a very angular or chiseled approach to

the surfaces can be completely appropriate if designing for an archetype that would appreciate such an aggressive, sporty statement. These shapes could express their narrative beautifully. Conversely, if one were seeking to convey a calm, refined, and relaxed impression, fit for a luxury statement, the sections and surface language would be very different. Added to this is the fact that there is a visual compositional dialogue between form, graphics, and their supporting architectural structure. For example, having a very rounded surface language combined with very fluid graphics may be too repetitive. The elements may have more harmony if they complement each other; fluid graphics with a more structured surface language, or vice versa. Hard versus soft, angular versus smooth, reflective versus matte, semi-gloss versus transparency... Sketching out one's ideas in the vehicle ideation phase of the process is very much the same as composing notes, melodies, and phrasing to a song.

## Explore Details, and Unique Selling Points and Features

With the architecture established, the basic composition blocked in, a complementary surface language identified, the last and

final step of the composition is to add the subordinate details. With vehicles, the finishing details can add a good amount of value to the visual equation. Many professional designers focus on them exclusively, such as lighting, components, and hardware. However, for the overview and high-level vehicle ideation exercise, it's only necessary to identify the key elements and give them enough definition and clarity to support the impact of the theme. In the front view, illumination can be key to communicating a vehicle's character. The old saying, "the eyes are the window to the soul" is very much the case. And as you can guess, that character and soul needs to be aligned with the archetype and narrative of the vehicle.

Other hardware and subordinate formal elements of the vehicle are also very important. Side view mirrors, for instance, add visual width and offer an opportunity to draw the viewer in. Door handles are a key touch point. One must consider not only how they look but also how they feel, and how they are placed on the vehicle to support the overall gesture and composition. Tail lamps offer a similar opportunity to headlamps but on the rear of the vehicle.

Many designers use them to add visual interest and intrigue to the rear end view of the composition. Finally, nomenclature, badging, and textures offer yet another opportunity to add a level of interest to the vehicle's aesthetic composition.

In terms of **aesthetic principles**, the objective is to imagine and illustrate the vehicle so the viewer reacts and reads it in three sequential phases. First read, second read, and third read.

*Architecture and proportion being the first read, graphic and surface language being the second read, textures and details being the third read.*

There should be a dominant main message, complemented by subordinate elements, followed by additional defining details. This three-stage process is a common way of delivering the aesthetic messages so they are complementary to each other, and, when taken altogether, each is in harmony with the visual narrative.

### Create a Range of Choices

Repeat the process of vehicle ideation adjusting compositional elements to create a bandwidth of solutions that address the predefined narrative in different ways. Human beings are hardwired to recognize subtle differences in visual relationships. The ideation process is about selecting compositions and relationships that suit a user's wants and needs in a way that is empathetic. To do this most effectively, one needs to examine different approaches. Subtle changes and alterations in compositional elements can yield vastly different effects. To create

the ideal, one needs to review and evaluate different choices. Having established these choices, one can then compare the strengths and weaknesses of each. Creating a bandwidth of ideas and solutions is part of doing due diligence for any problem-solving exercise. The easiest way to accomplish this is to repeat the steps in the vehicle ideation process but with three key objectives in mind: (1) Explore possibilities that shift and adjust architectural assumptions; (2) explore possibilities that shift formal and aesthetic assumptions; and (3) explore possibilities that shift and adjust both formal aesthetic and architectural assumptions. This will automatically yield at least four choices. Having established a bandwidth of solutions, one can then select the perfect combination to move forward.

△
**4.16** Jordan Meadows and Ken Nagasaka pictured with final model of Toyota Cross Cruiser concept
SOURCE: IMAGE CREDIT, JORDAN MEADOWS.

# Q&A

**FREEMAN THOMAS**
GLOBAL ADVANCED DESIGN
DIRECTOR, FORD MOTOR
COMPANY

*Freeman J. Thomas joined
Ford Motor Company in 2005
(Figure 4.17). He serves as
Global Advanced Design
Director, Ford Design. Thomas,
who is based in Southern
California, is responsible
for the company's advanced
design studios in London,
Dearborn, Irvine, and Shanghai.
After graduating in 1983
with a bachelor's degree in
transportation design from
ArtCenter College of Design
in Pasadena, Thomas began
his design career at Porsche
AG in Weissach, Germany,
working under Anatole Lapine
and Dick Soderberg. Of the
many Porsche projects he
worked on, being a part of
the 959 design team was a
highlight. In late 1990, he
was asked to join Audi AG
as chief designer, followed
by Volkswagen AG, where he
worked as chief designer under
Hartmut Warkuss. Thomas is
credited with co-creating the
VW Concept 1/New Beetle
that was presented in 1994,
and then led the iconic Audi
TT concepts in 1995. In 1999,
Thomas joined DaimlerChrysler
as Vice President, Advanced
Design Strategy and Vehicle
Architecture. In this role, he
led the team that created
the Noble American Sedan,
which became the iconic
new Chrysler 300. In 2002,
ArtCenter College of Design
awarded Thomas an honorary
doctorate for his work.*

△
**4.17** Freeman Thomas, Design Global Advanced
Design Director, Ford Motor Company
SOURCE: IMAGE CREDIT, FORD DESIGN.

*At Ford, Thomas has led
design teams on such
memorable futuristic show
cars as the Ford Airstream
Concept, Ford Interceptor
Concept, Ford Explorer
America Concept, Ford
Start Concept and Lincoln
C Concept. His teams have
also contributed to design
themes and strategies for the
new Ford F-150, the new Ford
Fusion and Mondeo, and the
new Ford Mustang. He plans
to retire at the end of 2017.*

**Question 1: Who are your
personal design heroes? Who
do you most admire, and why?
How does this key influence
relate to the work you do for
transportation and vehicle
design providers?**

▶ I have a few heroes in design.
The one that I most relate to is
Erwin Komenda. The reason
I relate to Erwin Komenda is
because I was first introduced to
his legacy when I was working
as a designer at Porsche. I
was inspired by the Auto Union

**76**

racecars from the 1930s, the Volkswagen Beetle, the Porsche 550 Spyder, the Porsche 356, and also in other areas of industrial design such as the Porsche tractor and Porsche four-cam motor. These were objects that I just fell in love with! And once I fell in love with these objects, I wanted to learn about how they were created. As I did more research, I kept coming across the name Erwin Komenda. So when I started working at Porsche, I kept asking questions about him. Many of the older modelers in the studio worked with him in the 1950s and 1960s and kept mentioning him. He was always true to the philosophy of form follows function, but there was always an endearing animated whimsical character to his lines. What you witness in nature.

And this is what introduced me early on to the idea of storytelling and character development. Without my really knowing about it, he created a foundation for me subliminally, and it was only later as I started to develop my own design language and the way that I looked at things philosophically that I saw these parallels. Then I was inspired to research him even more. Because he wasn't a designer or stylist, but Porsche's Chief Construction Engineer; he was the one who created not only the design and shape of the body, but also the chassis, the construction, and the holistic idea. Even when it came to working close with the other engineers, for example, in the development of the four-cam motor, he was responsible for creating the shape of the fan shroud for that motor that was functional, but again there is a whimsical charismatic quality to the design. There was a

sense of completeness to it. So he is one hero, and I could go on for hours about his designs and how they impacted me. How they've influenced me, and how he visually created honest stories.

My other hero is Frank Lloyd Wright. And the reason for him is because his iconic designs such as Fallingwater or the Guggenheim give an almost religious experience when you see them. They were and are still disruptive. The other thing that was really impactful about Frank Lloyd Wright was *when* he did them. He did Fallingwater in the 1930s! He was looking at cantilevered construction way before the technology existed: he had to come up with technology to support his ideas, and I always thought that was really interesting. For instance, Fallingwater, even though very advanced, doesn't have air-conditioning. The Kaufmann family wanted the place built as a weekend house or summer vacation house and envisioned it as something that would overlook the waterfall. And Frank Lloyd Wright looked at the site and said NO, it has to be ON the waterfall! That was an amazingly brave risk. He used the water flow to create a form of natural air conditioning. The stairway goes right to the river just before the fall, and the airflow caused by the water allows cool air to circulate through the house.

Another interesting characteristic of Frank Lloyd Wright was his full control of design and the customer. He wasn't so much designing for the client. In a way, he was affecting the client. He imposed his design philosophy so that the client would adopt his way

of thinking, the simplicity, and this edited way of life. And everything that was designed in his structures is an "ah-ha" moment, down to every detail. So you have Erwin Komenda, on one end, and you could say that he was a designer and engineer, and Frank Lloyd Wright as well. Now there is my third hero but I'll get to him in a bit ...

**Question 2: When generating new ideas for vehicles, how do you balance customer wants and needs with brand goals and objectives while ultimately delivering something new, fresh, and unexpected?**

▶ My thought process is probably less structured and more organic. It always has been. I know there are designers and researchers and engineers who want complete structure to how they go about things, but I tend to look at things organically and intuitively. And I always look at it by creating an original story. And sometimes it's just by being in the environment, being exposed day in, day out. I don't look at it from the standpoint of all of a sudden doing the research. Because for me, research is ongoing, it's observation, it's listening, it's not just looking at it from your own angle and brand goals. It's looking holistically at the way a brand is. To me, brand is a religion, it's a story. It is these things that make it worthy of being called a brand. Now stereotypically we look at brand as a name of a company. But brand transcends that. Brand is a country, it is a place, it represents objects, experiences and it's a state of mind almost. It's all of these various things that we believe in that are worthy of being called a brand.

Now how to deliver something fresh and new? Again I come back to creating an original story! It's about character development. A lot of times "fresh and new" are about giving the customer something they didn't know that they wanted. It's a bit like, if you listen to a customer they will tell you exactly how they want their meal made. But if you come in and say, hold on I'm going to let you try something new! They smell it, and they say, that smells good! What is it? And you say I'm not going to tell you what it is. I want you to just try it! In fact, I'm not even going to let you look at it! Then before you know it, it's something that they've never experienced! I think those are the types of experiences customers have that make them loyal to the brand. Because it's an original story, it's an original character development, it's an original experience, and it's something that can't be replaced. That's really the goal for a designer. It's to create that ultimate holistic experience. That goes away from just creating the façade like in architecture (going back to Frank Lloyd Wright) or with engineering such as Erwin Komenda, or for that matter, my third hero, Kelly Johnson.

Kelly Johnson led all the design and engineering for Lockheed Skunk Works. He was an aerospace engineer but had an amazing sense of aesthetic and intuition. And that's really the best kind of engineer. It's kind of like when you look at nature, nature is beautiful from the standpoint that form always follows function. It has evolved over millions of years to be what it is. If you take a river stone that has had water crashing against it for thousands of years and eventually forms into a naturally beautiful shape, it has a story to tell *why* it's that shape. So when I look at Kelly Johnson or look at his objects like the SR-71 Blackbird, it has this incredibly intimidating presence, but is also a symbol of future thinking. This is a great reference for a designer.

On the other side you look at a P-38 Lightning or a Lockheed Constellation and nothing talks about the romance of flight more so than these. Again, the foundation of the story is authenticity. Now today, if you look at an aircraft it's just an extrusion, because it's designed for cost and that's the best way to get two or three or four hundred people inside without changing the shape. But when cost is not an object, there is emotion. And a lot of times people are willing to pay the cost. And this is really a human response. So that's how I balance customer wants and needs. I start with their wants and desires. I want to know what is possible! Because I believe that what they need will be supported by the wants if it's done correctly.

**Question 3: Looking toward the future while considering zero emissions and autonomous technology, what are the biggest challenges facing designers in the vehicle ideation process? What is your advice to designers to conquer those challenges?**

▶  I think the technology is the easy part for an industrial designer because it's basically a creative packaging exercise. The most difficult challenge for a designer is going to be communication. When talking about communication, what I'm referring to is a couple of different things. One is communicating to leadership in a company about *why* they need to support this technology with unique platforms and unique packaging and investment. So that's on one side. The other part is the visual communication for why this technology needs to go beyond just normal research and focus groups. It has to be immersive; it has to come back to creating designs and forms that support the story. Let the engineering support the story and vice versa. And at the same time have it so endearing that you not only want/ desire it but you need it.

The other part that is going to be very important is communication to the customer. Again it's not unlike the previous question. It's so new, it's so fresh that the customer might even be slightly afraid of it because they've never seen it before ... We have a saying in the industry that we call "easy to love."

"Easy to love" can be dangerous because it only buys you a short time. It's reactive, like day trading. It really doesn't start the story from the original standpoint. I think things can be beautiful, they can be endearing, they can be functional, they can be socially conscious, they can be all of these things, but at the same time, the most important thing is when you fall in "love" with it! When you fall in love with something, you protect it. You defend it, and you will take care of it. There is nothing more important than what you care about. And I think that's when a normal appliance object transcends into something that we can say is alive.

**Question 4: What type of vehicle ideation strategies and techniques do you rely on to get the design team going when they've decided on a target user and have received a design brief?**

▶ For me, when I get a design brief, I question it. The first thing I instill in the designers is to question and to use their insights. Leverage their intuition and vision. On the one side, respect the target user and the assumptions given in the design brief. But don't take it as something that's written in concrete, because people make mistakes ...

I believe the role of a designer is to be a storyteller holistically. To create! You're born with insights, imagination, and vision. On the one side, I challenge them, because I believe that designers have got to become accountable. Accountability is where you've got the responsibility to create what the solution is to the problem. But also I believe they need room and I allow them to fail, because as they grow and time goes on, an experienced designer will slowly fail less, and succeed more.

It's also really about learning from each other. Not just coming from your own perspective, talent, and initiative, but being open. I believe in working loosely. It's important not to work too tight, especially in the beginning. When you're establishing the goals and when you first put the design brief up on the wall, ask how many different ways you can look at it. Ask how many different ways you can challenge it. Question the target user! Is that a moving target? Is the target going to be gone in the near term, or even in the long term? So you're building the story

to eventually have a plot, and you want to be sure that by the time the plot comes along everything surrounding it will still be relevant! Products, especially vehicles, are complex with many layers. An endearing and iconic product is like an onion that you can slice through all those layers to find an authentic and true center. If an object doesn't have this, the chances of failure are much higher.

**Notes**

1   Merriam-Webster. *Dictionary.* Available at: www.merriam-webster.com
2   Macey, Stuart and Wardle, Geoff. *H-point: The Fundamentals of Car Design and Packaging* (Los Angeles: ArtCenter College of Design, 2009).
3   Gibson, James J. *The Ecological Approach to Visual Perception* (New York: Taylor and Francis, 1986).
4   Norman, Don. *The Design of Everyday Things* (New York: Basic Books, 2002).

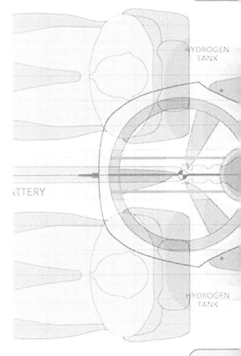

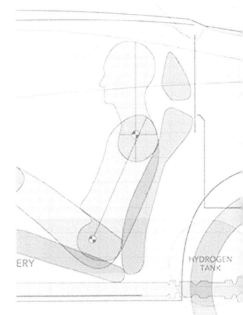

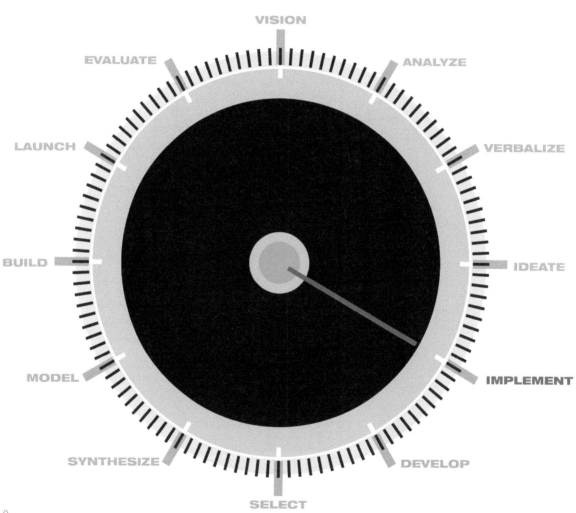

△
**5.0** Process locator gauge
Source: image credit, Jordan Meadows.

**IMPLEMENT**
SELECTING KEY DIRECTIONS

# CHAPTER 5

# IMPLEMENT
## Selecting Key Directions and Identifying Themes

## Selecting Key Directions and Identifying Themes

Design is a process of making calculated decisions in a sequential manner to arrive at a desired outcome. In the previous chapters of this book we have discussed the vehicle ideation process. Through sketching and visualization, we've arrived at the point in the journey where an idea has been given an image. Through a process of making subtle changes in the combination of the basic assumptions we can then generate multiple choices to form a bandwidth of ideas. Having a range of possible solutions is essential to giving the process due diligence. After all, design is about decisions, and decision-making is how strategy is "implemented" (see Figure 5.0).

In this chapter we will explore the thinking that goes into narrowing the bandwidth and identifying the key themes to move forward. In most cases the amount of choice in the beginning of a program is widest with a broad range of possibilities. This is known as a development funnel. In implementing a plan of attack or **design strategy**, one can then begin the process of narrowing

the funnel (Figure 5.1). This is the thrilling point when a designer can begin to get excited about how cool the vehicle might actually be. It's also the stage where strengthening the propositions, double-checking the competitive landscape, and focusing the design strategy become crucial.

## Design Strategy: A Brief Overview

To a certain degree, this entire book is about design strategy and its many phases of implementation. However, this particular point in the process is a good stage to highlight the subject and its importance. The design strategy is effectively the actualization of one's creative agenda. It is often easy to confuse it with a **mission statement** or **design brief**, both of which frame the objective. The design strategy, however, is how to approach the decision-making in generating and deciding on solutions. For many people in the creative field, it's possibly when the head and heart truly start working together in unison.

Design strategy involves addressing objectives that are crucial to moving a program

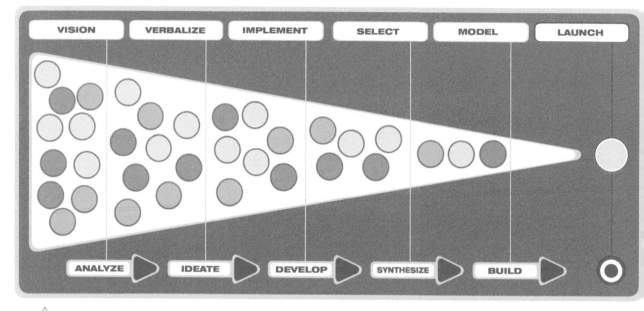

VISION | VERBALIZE | IMPLEMENT | SELECT | MODEL | LAUNCH

ANALYZE | IDEATE | DEVELOP | SYNTHESIZE | BUILD

△
**5.1** Info graphic depicting typical product development funnel in that many ideas and proposals are filtered and distilled down to one. The design strategy, in conjunction with the objectives of each stage, guides and informs the selection process.
SOURCE: IMAGE CREDIT, JORDAN MEADOWS.

forward. When implementing the brief, it helps prioritize the most important questions that need to be asked. These are a direct result of the organization's business strategy. It's the moment that a designer can begin to generate or evolve a visual **DNA** and **aesthetic principles**. So, for example, after a range of ideas have been sketched out, one can then ask which ones are most appropriate to the brand from an aesthetic point of view. Which ones have the most family resemblance to the pre-existing vehicles created by that manufacturer? Which ones feel the freshest and most inventive versus predictable or evolutionary?

Implementing a design strategy can also foster the adoption of technology and promote innovation. When a range of ideas has been established,

it's often the outlier that could hold the most potential for the brand. Because design strategy informs all the touch points in a vehicle's design, it effectively translates into actionable solutions governing how to address technical challenges. It highlights the opportunities to differentiate. Form can follow function OR form can follow fun! Design strategy resulting from a strong **narrative** informs how that choice can be made.

Finally, design strategy, when implemented in conjunction with the brand DNA, can inform the order in which products and services can be launched. For example, once a bandwidth of ideas has been sketched out, it's quite often the case that some solutions are more near-term, having relevance for today versus other solutions that are

far-reaching and could take some time to adjust to from an aesthetic or technical standpoint. In this regard, the design strategy is a crucial part of the long-range plan for a brand. Vehicle manufacturers quite often are very focused on a particular user. Having a clear vision of how a design strategy could evolve can illuminate the way to endearing a brand to new customers. For example, Mercedes-Benz in the past was acknowledged as a venerable manufacturer of mature luxury cars. Understanding this caused them to evolve their design strategy to become much sportier and target a younger buyer. This shift in design strategy governed the selection of the sketches, models, and themes they chose to pursue. The result currently is a range of cars that are sportier and more dynamic than they've ever been. Having a design strategy is

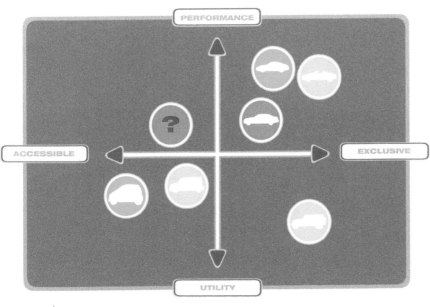

△
**5.2** Info graphic depicting an example of a typical positioning chart. Design strategy and product planning are used to assess the vacant areas with intent to inform a theme selection process.
SOURCE: IMAGE CREDIT, JORDAN MEADOWS.

essential to understanding existing problems, ongoing challenges, and enabling a group to achieve their long-term goals.

### Understanding Segmentation and Competitive Benchmarking

Another consideration in weighing choices and making decisions involves understanding the target segment. This requires a certain degree of **competitive benchmarking** and overlaps the field of product planning and positioning (see Figure 5.2). While most designers are not typically responsible for this, it's important that they have an understanding of what it implies. Most **vehicle segments** have a core with key players. For example, the Ford Fusion is a popular offering in the midsize CD segment. When developing a replacement, the main competitors from companies such as Toyota and Honda and

General Motors are benchmarked for appearance and functional attributes. To some degree this helps establish the strengths and weaknesses of themes created for the next generation Ford.

Anticipating how competitors may evolve and predicting their future position within the segment is an important part of product strategy and identifying key themes to move forward. Forecasting what may look cool from an aesthetic point of view, along with unique selling points for packaging and functionality, is critical. This is based on good quantitative research, an understanding of market trends, and a fair degree of intuition and qualitative gut feel. These traits all apply to selecting key themes to move forward in the design process. It's an exciting point where a future classic can be given its first consideration.

### Magic, Risk and Balancing Science with Sorcery

Because no one can truly predict the future, it's important to have a good range of possible solutions that fit the brief, align with the design strategy, and reflect the product narrative. One way to do this is to identify at least three different directions, giving coverage from complementary yet opposing perspectives. In selecting key themes to move forward, one of these directions should include a proposal that is market-focused and oriented toward what is certain to do well against existing players, given current assumptions. The second theme, however, should honor the purest vision, the sentiment of the brand, and be designer-focused. Finally, the third should be a disruptive outlier derived from a radical shift in future assumptions.

It is important to note that all three of these must have relevance to the user. The first two will admittedly be more comfortable and address the pre-existing business goals in a direct way. The third, however, should feel unexpected and challenging. This will be the proposal that pushes the boundaries. In embarking on any creative journey, one must be prepared to make discoveries. Many discoveries by nature are unexpected. For this reason, it's crucial to embrace tension and the unknown. It is these exploratory ideas that are often uncomfortable in the beginning, yet prove to be the most informative and rewarding. In the end, it's the one that may turn out to be pure magic!

# Q&A

**PETER SCHREYER**
CHIEF DESIGN OFFICER,
KIA/HYUNDAI

*Peter Schreyer, President and Chief Design Officer (CDO) of Hyundai Motor Group, is responsible for the design activities for the company's Kia, Hyundai, and Genesis brands (Figure 5.3). Schreyer's automotive career began in 1980 with Audi, and his 25 years at the Volkswagen Group saw him successfully participate in a wide range of projects, becoming one of the driving forces behind the image change of the Audi brand.*

*Schreyer has received numerous recognitions and honors throughout his career, including "Design team of the year" in 1999 for Audi. In 2003, he also received the highest personal recognition for his unique design talent: the Design Award of the Federal Republic of Germany.*

*Schreyer was born in Bad Reichenhall, Germany, and grew up on his parents' farm, developing a passion for cars and airplanes. He studied industrial design at the University of Applied Sciences in Munich, followed by a course in transportation design at the Royal College of Art in London. He also received an honorary doctorate from the Royal College of Art in recognition of his contributions to automotive design.*

△
**5.3** Peter Schreyer, President and Chief Design Officer (CDO), Hyundai Motor Group
SOURCE: IMAGE CREDIT, HYUNDAI KIA.

**Meadows: Regarding the Kia Soul, what made you guys go about an MPV in such a cool way?**
**Schreyer:** To go back to when the first version of the car was launched, it was created as a show car; the first Soul was actually developed as a concept. And the concept was actually very close to what the production wound up being.

It was more of an SUV or CUV type of car that came from design and it started as a show car presented to the top management. They were so excited about the show car that everyone rallied around the idea and said let's do it! And this is how it came into production.

At that time the company was looking to set themselves apart and it was just around the time that I arrived. I was only a part of the development of the production car about halfway through the process of the first Soul. The top management wanted to set

themselves apart from other companies through design and they really wanted the production car to be like the show car. We also went through a lengthy discussion about the name because there were some concerns about the name in some regions and how the word translated into different languages was something that we considered (Soul).

I also feel that the name has something to do with the success of the car. Some people think that the name doesn't matter but for me it makes a big difference. I remember back at Audi and working on the TT project; for a time, some in the organization considered calling it an A3 Coupe!

In the end, the name is such a part of the character and it helped guide the entire project. I think when the name tells a story and fits with the story of the car, somehow it becomes meaningful. And when the combination is right it makes a big difference.

**Meadows: What's interesting about the Kia Soul is that it has such a strong graphic statement on the road (see Figure 5.4). The wrap-around glass with a blacked-out pillar. A flat roof, minimal wedge stance combined with a strong C-pillar. It is very memorable, and has quite a following. (see Figure 5.5).**
Schreyer: It is quite a strong statement. It is also one of those cars where maybe some people hate it but then there are a lot of people who love it! This is the thing that happens with a car that has character. And that friction can make the car last longer, more relevant, and get people to talk about it. They discuss it (see Figure 5.6).

△ △ △
**5.4** Kia Soul concept, theme sketch
SOURCE: IMAGE CREDIT, KIA MOTOR CORPORATION.

△ △
**5.5** Kia Soul, second-generation
SOURCE: IMAGE CREDIT, KIA MOTOR CORPORATION.

△
**5.6** Kia Soul, second-generation, front three-quarter
SOURCE: IMAGE CREDIT, KIA MOTOR CORPORATION.

**Meadows: In your career history, is this something that you've always tried to push for?**

Schreyer: I think maybe I do. For me it's always important, as I said before, that a car has a story, if possible, everything that we do should have this. This is also important for the designers, if it's just about some lines up or down forward or back: sometimes I have to tell the guys (my designers), I just don't care. I care about the proportion, and I care about what expression the car has. I need to recognize it when you see from a distance whether it has a line on the side or not, well, whatever ... This of course is part of what we do but it's not at the core of what we do. The question is, does the car have tension and does it have a forward movement? Is it aggressive or laid-back and relaxed? What kind of statement does a project need to make?

**Meadows: Looking back on which project did you begin to develop this idea of storytelling, or was it something that you learned from someone else, a leader or mentor figure, perhaps Mr. Warkus?[1]**

Schreyer: I think this is something that maybe came from working together and from learning from several people. Mr. Warkus was one of them. I remember when we first started working on the very first Audi Avant. He came in and said we need to do a wagon. Remember, at this time wagons weren't very desirable. They were usually for people who had a business, a baker, or butcher, etc. They were for someone who needed to deliver things and he basically said he wanted to do a wagon where not only a decorator could put their materials and things in the back but where you

could also put skis and golf clubs and you could take an exciting holiday perhaps. And this is how the story or idea of an "Avant" came along. This storyline is also how Avants became a lifestyle vehicle. This is also how wagons transformed and became so successful.

**Meadows: Do you remember what year it was?**

Schreyer: I remember exactly. 1978! That was the year that I went to start my internship!

**Meadows: There is still somehow a difference between a wagon and an "Avant," somehow what you guys did back then made that silhouette very desirable. With the exception of the Dodge Magnum, or Cadillac CTSV, that type of vehicle actually never really caught on in mainstream car design in America. I suppose the marketplace just feels differently about them here in the United States.**

Schreyer: Yes, I've waited for them to get traction for a while now. If you still want style and functionality. But prefer an alternative to an SUV, an Avant or some sort of cool station wagon is such a great machine. The Dodge Magnum you guys did at Chrysler was a great take on the idea with a cool name as well to fit that story.

**Meadows: What made you begin your career with Audi?**

Schreyer: It was a bit of a happy and lucky coincidence. I was at school in Munich for product design and industrial design. Cars were always something that attracted me, but for me the car industry was a bit of a different world—very distant and not accessible. We had to

do an internship back then to complete our studies and one of my instructors worked in interiors at Audi so it kind of came together that I got an internship. Things were very different back then, and the image of Audi was very different than it is now.

**Meadows: Not quite as sexy?**

Schreyer: Not at all, I still remember the moment I walked in the door and saw a hard model of the original Audi Quattro! This was the first time I had seen a design model. I remember it was painted in white like a racecar, and then they told me that this car was going to have over 200 hp!!! I thought to myself how cool is that! It was all very top secret at the time, there were very impressive designers working there. Martin Smith was there, Peter Birtwistle and John Heffernan. They, along with some others like Gerhard Pfefferle, were basically the core team. The internship was an experience that basically changed my life. Bertie and John Heffernan and Martin Smith knew about the Royal College and Martin Smith took the time to speak to Mr. Warkus and he basically initiated and helped to convince Mr. Warkus to send me to the Royal College. I still remember him asking me if I was interested in going to the Royal College of Art. OF COURSE I WAS! Before my internship I didn't even know the place existed but it sounded very impressive! It was the most prestigious college in the world.

And it was all very exciting, so I went and I'm still grateful to this day that they gave me that sponsorship and that opportunity. I always felt grateful for them that they believed in me, and after that there was no question of any other

IMPLEMENT
SELECTING KEY DIRECTIONS

company, Audi was the place for me so of course I stayed for a long time.

Meadows: **I think in a way that path you took was very influential for me personally because I also did product design and I wanted to be an automotive designer. In the design community and at my school, there was always a philosophical divide between car designers and product designers. The place where designers managed to span and bridge that divide was the Royal College. During my undergrad work at RISD, I remember reading about how you went there and many other influential design chiefs also attended. I also think the product design approach is very much apparent in the work that you guys did at Audi. At that point in time, did the company have a strong strategy from a design standpoint?**

Schreyer: After the RCA I returned to Ingolstadt and started in exteriors. I think, yes, at this time Audi had a strategy but still had a very small product range. It was just the Audi 80 and Audi 100 and both had a wagon variant and that was about it. We also had the Quattro, that was very special. At that time in the mid-1980s it was still very much about technology: "Vorsprung durch Technik" was the slogan. We were very much about aerodynamics, aluminum, lightweight, and a fun sporting character. But at that time we were challenged with an image that wasn't as desirable as BMW or Mercedes, so we did a lot of research on what our strategy could be for differentiation, and to make our mark. A lot of customers felt that even though the cars

were superior technically, they just didn't have an emotional draw. Some customers used to say that an Audi had the character of a German shepherd or even a police dog. Very reliable, and can do a lot of different things well, but you just wouldn't want to cuddle it. At that time, we had a very special group of designers who also had a feeling for what the brand needed. It was a super group of people who have gone on to be hugely influential. We somehow had a similar view and common sense or feeling amongst us all of what needed to happen. In a way this was all due to Mr. Warkus selecting the right people who had this common vision and shared sensibility.

Meadows: **So, in a sense, much of the way you give direction or implementation strategy is putting the right designers together? Rather than prescribing and saying, "Guys, go out to do this," a lot of it is finding the right personalities and putting them together and letting them discover on their own?**

Schreyer: Yes, I often wonder if you can do this as a strategy, or if it's a matter of happy coincidence. Or somehow the right people come together at the right place and time. In any case, I do think a bit of this is actually happening at Kia and Hyundai right now.

Meadows: **There are very few people who would argue with the success of Kia and Hyundai from a business and also a design standpoint. Much of the success has happened under your guidance in the last 10 years. Since then it's been steadily growing in significance. By every measure, the company**

has a level of respect now that it didn't actually have back in 2006. What cars would you say were the most significant in terms of earning that respect?

Schreyer: From the Kia side, it was definitely the Optima and the Sportage and the Soul. But overall the entire line of cars earned a lot of respect. I think we also managed to have great success not only on the exterior but on the interior side also, so the customers viewed the cars more holistically. Hyundai was a bit of a different story because I was not directly involved in a lot of the work there until recently. But I will say what they've done with the fluidic sculpture design language was quite a daring move for a company that was not established like a Mercedes or BMW. For example, if they had done such a bold move, everyone would've said, "Sure; looks great." But for Hyundai at the end of the day it was a very daring move. Of course, in this game, no guts, no glory, as the saying goes. And they've made it work and it has made a big difference for the brand. The first fluid Sonata was quite extreme for its time but in a way it paved the ground for a lot of other companies. In this case a lot of other companies have actually been following the car's look and appearance, and this is maybe the first time where the reputation of a Korean car company became established as a leader.

Meadows: **Bold moves for sure, and I think the amount of time or the speed that it takes you guys to create and bring your vehicles to production is also pretty amazing.**

Schreyer: Yes, that's correct, the speed of the company is quite remarkable. I see it in many of the projects that we are doing even now. We will have an initial presentation and then literally a few months later we are looking at finished hard models and the content is really amazing. It's all there ready for production, proper prototypes with full design intent. I can tell you one example of the Genesis G90, our large luxury sedan. The project began with quarter scale models that I selected when I started at Hyundai in my official leadership role. It will be three years to date in January 2016 and the car will debut in about a month's time on the market. So from the quarter scale model development it was about a 36-month process from the start (see Figures 5.7–5.9).[2]

**Meadows: At Hyundai and Kia, how are the selections made from different proposals? How do you and the other executives decide what design direction to take and which proposal to move forward?**
Schreyer: Very often we have a process that is kind of a global competition amongst all the design studios. We have four models in the beginning and then pretty soon we reduce it down to two, and very quickly reduce it down to one and go from there. The decision-making is very quick and sometimes it is indeed quite stressful for us. Sometimes we will need an alternative but generally we have to be quite sharp in our decisions.

**Meadows: And it relies mainly on intuition, your experience, and your understanding of the marketplace and customers?**
Schreyer: Yes, of course. And then we also have presentations that take place in Korea. We get feedback from everyone because that's where everyone

△△
**5.7** Hyundai Genesis, featuring signature front design DNA
SOURCE: IMAGE CREDIT, HYUNDAI MOTOR CORPORATION.

△
**5.8** Hyundai Vision G concept, interior
SOURCE: IMAGE CREDIT, HYUNDAI MOTOR CORPORATION.

comes together and makes an agreement. This is quite good because the chief designers and designers are involved. We all meet and get together from the different studios that are also in competition. Though we're spread out around the globe, it gives us a chance to have a common sense and common understanding.

△
**5.9** Hyundai Vision G concept, exterior side view, car created to preview the Corporation's design language for premium Genesis brand, debuted at the Pebble Beach Concourse d'Élégance, August 2015
SOURCE: IMAGE CREDIT, HYUNDAI MOTOR CORPORATION.

And also feel as if we belong to the same team and are moving forward with our decisions together.

The main studios are in Germany, California, and of course our headquarters is in Korea. There is also a China operation, along with a research group in India and other contributing offices. The organization is quite widespread with a large footprint. Some of our regions also have individual R&D centers; for example, we have a technical center in Europe where they make adjustments for the European market. We also have a test center at the Nürburgring that gives us the opportunity to do research on high performance vehicles.

**Meadows: Do you have rotations and designers moving among studios?**
**Schreyer:** We have designers moving from Europe to Korea and vice versa and then also from California to Korea and vice versa, but very few from Germany or Europe to California. I would not mind it but there are some challenges with visas and work permissions in that regard.

**Meadows: I found personally that the creative chemistry in most car design studios is like a see-saw; effectively a balancing act between competition on one side and collaboration on the other. Because designers are naturally competitive and they also have a good dose of insecurity, they will usually try to show that they can do a better job than a counterpart. What**

end of the spectrum would you say that your team works on more? Leaning more toward the collaboration method or are they more competitive with one another?

Schreyer: I think it's like almost every other studio, every single person always wants to win. But I do think within the studios they have a good atmosphere and people do feel like a team. When something is chosen and is selected for production, they can all stand by it. It comes from their group as a whole so everyone feels very proud of it.

Meadows: In some organizations you can point to one individual and say he or she did that car. But in other organizations one can't say this because it is very much a team effort. The dynamic is similar to sports. Some are very much about individuals like tennis or golf, and others are about teams like baseball or football.

Schreyer: Yes, it also sometimes depends on the project. Like in exterior studios, for example, you have one designer who did a quarter-scale, and then a different one had to follow it up into production. There are others involved like the team leaders and directors or people like myself who are helping and supporting the creative process. Sometimes there is a more experienced designer joining in for guidance, so finally the ownership lies with several people in the team, but you can trace it back to one individual. The final car is the result of everyone working well together. But many times it traces back to an individual who executed the sketch. (The key sketch!)

This is something that is very important and we need to keep: The key idea that leads to a model and is the target. Often we will have a wall full of sketches but we need this key sketch to refer back to in the process; for example, sometimes designers will have lots and lots of sketches and will have a model as well but when you look at the clay—the model is not the sketch. They haven't translated it ... Something is missing ...

Sometimes you can even look at a rear three-quarter sketch and you can see the character of the window graphic. The attitude of the side view, etc., it promises you something and gives an impression of what the front might be even though it's a rear view. The viewer has a vision or wish for what's on the front.

Meadows: It's almost like when you see the silhouette of a person at night or from the side you think of what their face might look like?

Schreyer: Yes, and sometimes it is a disappointment. This is something that we also learned in the early days from Warkus: (*You look at the sketch and try to capture it!*) Sometimes I feel a bit disappointed that designers now work too much with Photoshop or they'll take something that's pre-existing and they start changing and shifting it around. And I'll say, "Okay, where is your idea?" Whereas if you do a sketch, it still has your handwriting and feeling to it. The same is true with a tape drawing. Perhaps it is a bit old school and I might be a dinosaur but at the moment we're losing the culture of taping on the model. Often I'll talk with a designer and will put a line on the car and say, "What do you think?"

This part should change here or there ... I tape it on and we agree on a change. He'll say that he sees that the line needs tension and what needs to change, but will want to make a change in the computer. The problem is, when you transfer back into digital mode so much is lost …

Cars are sculpture—you have to touch them! You have to walk around them. Now, of course, there are time constraints and resource constraints and you have to move the data back and forth so it can become tedious, but if you sketch it with tape on the model, and if you can do a rough tape sketch in 3-D on the clay, you can read it for yourself and estimate how much needs to come off the model and how much needs to be put on. It's real time and you get the relationship to real life and the body rather than guessing at a screen. Ultimately a car is a 3-D thing that you'll see on the street and you have to have an understanding of it. Now the process has changed and evolved and some are able to translate their ideas into 3-D data very early on.

Meadows: When you're selecting a sketch, are you looking at what will make for a good model or do you look for an image that has a bit of magic on its own?

Schreyer: Both.

When you look at a sketch, you cannot avoid the impression that it gives you. You can't look at it like a computer and just analyze what line goes where. A good sketch has that bit of imagination and allows the viewer's imagination to start. When looking at it you say to yourself, "I would try this and

I would do that ..." You complete it in your mind, and this is what I want to try to communicate to my designers.

Often they will do the sketch but they don't see the same things that the viewer does. Sometimes I get a presentation and the designers will show the side view, the rear three-quarter, the front three-quarter, etc., along with the final idea, fully rendered. And then I'll go to the little small doodle or thumbnail and I say, "Look at this one!! It's here! The answer is here!"

It's almost like when they do the final image and they become stuck or paralyzed and it's completely lost the character of the sketch. BUT the key sketch has it!

**Meadows: Something about the magic of the moment when the hand moves across the page that makes it special?**
**Schreyer:** Exactly!

**Notes**

1   Hartmut Warkus, former design executive for the VW Group and longtime mentor to many VW Group design chiefs.
2   Figures depict key design direction for Genesis brand as the G90 production car images were secret at the time of this interview.

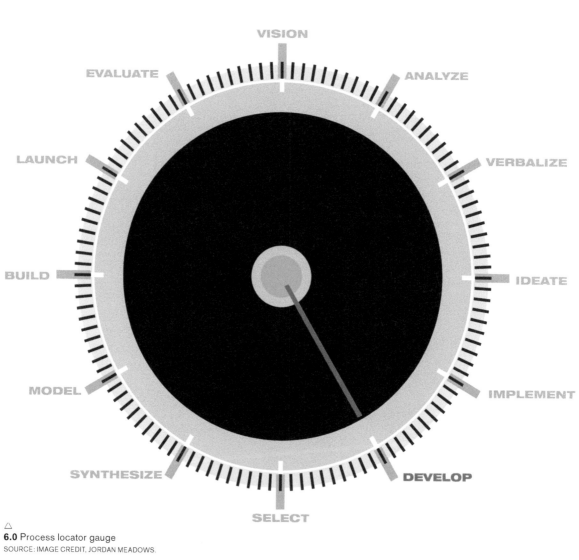

VISION

ANALYZE

EVALUATE

VERBALIZE

LAUNCH

BUILD

IDEATE

MODEL

IMPLEMENT

SYNTHESIZE

DEVELOP

SELECT

△
**6.0** Process locator gauge
SOURCE: IMAGE CREDIT, JORDAN MEADOWS.

# CHAPTER 6

# DEVELOP

## Developing Key Themes:
## Ford Mustang Case Study, Part 1

In Chapter 5 we talked about how selecting key directions and identifying key themes are important in the process. We learned through an exclusive interview with Hyundai and Kia Chief Design Officer Peter Schreyer how this is done. Once we've given ourselves several choices to choose from, we can narrow down to the strongest proposed directions. Many factors can contribute to the selection process but ultimately it's governed by a **design strategy** for that individual program.

Once the key themes have been selected, one can then begin the important process of refining and (developing) those key themes (see Figure 6.0). In Chapter 6 we will explore how this is done using the example of the 2015 Ford Mustang. This project is a prime example of how key themes can be refined and developed because quite frankly a sporty coupe is the type of vehicle that almost everyone has some familiarity with. The discussion and focus can then shift away from the type of product and its attributes to the specific design execution and (development) of the selected themes.

||||||||||||||||||||||||||||||||||||||||

# Ford Mustang Exterior
## Part 1

△
**6.1** Mustang silver side view of 2010 model and basis for design development. It was agreed that the replacement would be lower, wider, and more dynamic in almost every respect.
SOURCE: IMAGE CREDIT, FORD DESIGN.

The word iconic can be an overused term especially with regard to some automobiles, but once in a while the description actually fits. There are few automobiles with historical and cultural significance to rival the Ford Mustang. Tackling this subject matter alone would be a very daunting task for an individual designer to take on. Fortunately, the process of designing any car is almost always a team effort. In the example of the 2015 Ford Mustang, this was very much the case! The final car was a seamless blend of the efforts of many talented team members. On that point, I'm proud to say I contributed to the program along with Tyler Blake and Kemal Curic as one of the three key exterior designers. However, the entire design team of Ford North America helped to deliver the sensational piece of

design work that is now on the road. The Mustang is not only a car but also a cultural artifact. It is a true piece of American pop culture. This case study tells how the design team updated and re-imagined a modern classic using the process of refining and developing key themes.

**Releasing the Wild Horse**
In early 2009, discussions began to take place at Ford about how the design team would tackle the 50th anniversary of one of the most popular cars the corporation had ever created. Ford Motor Company had done a remarkable job of outlining core vehicles that would deliver financial success for the company by selling in volumes of upwards of 2 million units annually. Though these core products are essential, one cannot mention Ford without also thinking

Ford Mustang. The project would debut at the 50th anniversary of the brand. So everyone knew it had to be nothing less than amazing. On top of that, the project held a special place in the hearts of our senior management with the Ford family included—no pressure whatsoever, right? It was clear in these discussions that the project was going to be passion-based, and elicit the A+ game from all of the team members.

## Character Development and Processing Imagery

Mustang was beautiful because the car perfectly reflected the archetype ideals of the Sexy Rebel. It represented this fist smashing through a glass plane. Breaking conventions, and leaving an impression. It also had to be cool. Not just any type of cool. Not like a rapper or a hipster, or someone contemporary, but very much like Steve McQueen. That Steve McQueen reserved, masculine, cool. Imagine McQueen's heavy brow with calm eyes and strong, chiseled jaw. That's Mustang. That's the Sexy Rebel. The face of the vehicle, with its heavy brow and reserved masculine cool, confronts the air in a very intentional way. Vertical and flat, it busts through the air, like a fist. I would even argue the taillights with the signature verticals are like knuckles. They leave an impression on the viewer and in the air. And again, this is the Sexy Rebel, wanting to confront and leave a mark. With Mustang, it's so clear that a vehicle moves through the world with great intention and communicates a driver's aspirations. And how you move, and what kind of impression you want to leave, literally, becomes very important when determining the shape of the vehicle. Are you surfing through life? Are you slicing through the air? Mustang doesn't do either: Mustang confronts and leaves an impression, and how beautiful is that! So understanding the narrative and aspirational motivations and goals revolving around the customer is key. It's not just about the numbers or the rational functionality. The vehicle is a symbol of how you want to move through the world and realize, or reaffirm, your ideals. This is the thing that is so unique and beautiful about cars. They embody the poetry of movement, a type of ballet. And they can reflect our deepest dreams and aspirations.

(Angela Weltman, cognitive psychologist, on researching Mustang)

In Chapter 2 of this book we discussed the importance of analyzing and understanding the wants and needs of the customer. With regard to the Mustang, this was even more important as Mustang owners and customers tend to be very passionate about the car that they purchased. Fortunately, the team was composed of enthusiasts. In a sense we were designing a car for ourselves or at least something that we would always want to have. This made the process of engaging customers to elicit their insights and wants and needs for a new Mustang relatively easy. On the one hand, the Mustang is a car created by enthusiasts for enthusiasts. But we also had the ambition to expose new customers and buyers to the wonder of the brand so they were also researched and interviewed in great detail.

So it became clear very early on what the car had to be: fast, fun, and attainable, ensuring a sense of freedom, expression, and individualism.

Coming up with keywords that an entire organization can rally around is very important in the development process. The keywords that defined the 2015 Mustang were:

1. Breaking Out
2. Masculine Cool
3. Power and Control

Automobiles are unique in that they portray a character. When you develop an automobile, it's sometimes helpful to imagine and visualize a story that the character is involved in. We've explored these ideas in the first few chapters of this book. From a historical standpoint, many on the team always viewed the car as a perfect fit for a Steve McQueen character. Cool, sophisticated, with a masculine confident persona. If you're a car lover and a movie lover such as myself, how could you ever forget the 1968 classic *Bullitt*? In fact, looking at the car, one can almost hear the score from Lalo Schifrin playing as it cruises by.

It had been decided by the program development team quite early on that the next generation Mustang would not just simply be a better **pony car**. From an engineering standpoint, the group wanted to focus on benchmarks that were at least a class above; for example, the Porsche 911 and other international competitors. It would be lighter, efficient and more intelligent from a technical perspective. So, in turn, looking forward we hoped that the

Mustang would evolve to fit a modern hero like a Ryan Gosling or Brad Pitt character. All three of these actors became quite popular for being in action films and are known for a unique image. They balance their sexy leading Hollywood male persona with a sense of intelligence, sophistication, and a mysterious quality that makes them enigmatic. And while they have an international appeal for these values, they are also considered uniquely American.

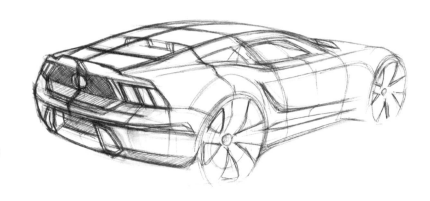

The 50th anniversary Mustang would be packed with technology that would make it class-leading and groundbreaking in many respects. It would also be one of the first Mustangs to go international. In short, it would be one of the most technically advanced cars that Ford had put to market. So much of the imagery that was discussed and developed and proposed for the exterior of the car actually jelled with the engineering that was being developed within the car.

At this point the team had a very clear understanding of the voice of the customer and had assigned some relevant imagery to the program, so they could then begin the process of shaping the **silhouette** and architecture of the vehicle.

### Establishing an Architectural and Visual Foundation

Vehicle packaging and **vehicle architecture** are at the core of any program. Establishing the major compositional elements will create a silhouette and a centerline that can evoke a mood in character as defined in the stage we just discussed. It will also establish the basics of

the user experience. Through intensive customer research with enthusiasts, everyone understood that the Mustang had to have a long hood, short deck, and low roof. An aggressive stance was a key to success. At the same time there was a goal to make the car lighter and more efficient than its predecessor (see Figure 6.1).

This meant that the frontal area was going to be reduced so it could perform in a more aerodynamically efficient way. The team also decided that the car needed to be significantly lighter to perform better. To evaluate these moves, the team generated proportional studies where the existing design was morphed to a new proportion and package so that the challenges could be understood in the changes evaluated.

At the same time, by mid-2010, the designers began the process of sketch work developing the look and feel of how the next generation Mustang would appear (Figures 6.2–6.5). At this point in time, design management had done a fantastic job of creating consistency across Ford's vehicle line. This was due to the fact that most of the cars had a family resemblance to the front-end appearance. Central to this idea was a dominant trapezoidal opening with horizontal lighting elements expressing precision and technology.

A key decision was made at a high level that the next Mustang would have a contemporary face that would convey a sense of masculine cool, power, and control, rather than evoking heritage the way the 2005 Mustang did. The various design proposals that we produced were all derived from this basic direction: to lose a front bumper shelf and depart from something that every Mustang design had done since 1964 (see Figures 6.6 and 6.7). In a sense we all knew what type of face the car had to have; it had to be connected to Ford's overall family, yet portray

◁◁◁◁
**6.2** Mustang early-stage rear view themes sketch, indicating trademark try bar tail lamp and tapering canopy
SOURCE: IMAGE CREDIT, FORD DESIGN.

◁◁◁
**6.3** Mustang rear early-stage study exploring an alternate execution for rear graphics and single frame side glass graphic
SOURCE: IMAGE CREDIT, FORD DESIGN.

◁◁
**6.4** An early-stage Mustang rear investigation exploring an alternate rear and composition
SOURCE: IMAGE CREDIT, FORD DESIGN.

◁
**6.5** Top view Mustang design themes sketch, indicating deep rear haunch and progressive window graphics
SOURCE: IMAGE CREDIT, FORD DESIGN.

△ △
**6.6** Front three-quarter design image displaying Mustang character in a progressive way, yet without a typical front bumper offset
SOURCE: IMAGE CREDIT, FORD DESIGN.

△
**6.7** Influential Mustang front three-quarter design image expressing front-end character
SOURCE: IMAGE CREDIT, FORD DESIGN.

△△
**6.8** Early-stage Mustang rear three-quarter image exploring alternate rear and graphic
SOURCE: IMAGE CREDIT, FORD DESIGN.

△
**6.9** Early-stage Mustang front three-quarter image exploring dominant central trapezoidal opening
SOURCE: IMAGE CREDIT, FORD DESIGN.

studio. Moving through the body side, one of the proposals featured a cleanly smooth A-line that ran the length of the car and anchored the side graphic. The aeronautical roofline was a key feature the design management wanted to play as a link to the original '64. In the C-pillar area, it included slots and vents to also hearken back to the original (see Figures 6.10–6.12). This was complemented by the traditional hockey-stick element in the lower body-side, another long-standing Mustang cue. The modern and progressive feeling of the proposal came from the sculpted sheet metal that was stretched in a taught fashion between points of interest. Other body side compositions maintained a traditional haunch over the rear wheel.

As the proposals took shape, it quickly became evident that this was a very different type of animal that we were designing. Not just because the design work was different, or pretty darn good in our collective opinion, but the package and architecture of the vehicle were in fact substantially different from the outgoing 2010 model. The car was lower, wider, and more aggressive in terms of how the layout was constructed. The basic engineering group had done a fantastic job, evaluating all of the competitors and assembling the basic architecture and package so that not only would it be best in class, it would be truly inspirational from a design perspective. It was almost as if the outgoing Mustang had been to the gym and done lots of cardio and weights. So whatever means by which we as designers were going to dress the engineering package, it was bound to look more athletic, lean, agile, and fit for purpose.

a uniquely Mustang character. To honor the strategy, most of the proposals that were created had a very similar front-end. Through the body side, roof, and rear, the designers had the task of interpreting Mustang's signature elements and delivering them in a fresh and compelling way. As stated, the front-end was largely prescribed through each designer interpreting the subject matter from their own unique point of view. Hundreds

of sketch proposals were done and submitted from Ford's various studios around the globe. Many were hyper-modern and futuristic. Others were more contemporary and some heritage-based (see Figures 6.8 and 6.9). We all wanted to convey a sense of power through large intakes and perhaps hood scoops. Through successive editing reviews, the proposals were pared down to three, coming mainly from the Irvine studio and the Dearborn

△△△
**6.10** Mustang front three-quarter image depicting dominant front trapezoidal opening that also explores headlamp graphics leaning forward in traditional Mustang fashion
SOURCE: IMAGE CREDIT, FORD DESIGN.

△△
**6.11** Early-stage Mustang design image exploring an aeronautical roofline hearkening back to early generation Mustang fastbacks
SOURCE: IMAGE CREDIT, FORD DESIGN.

△
**6.12** An early-stage Mustang side view design image exploring a linear body side treatment executed without a rear fender haunch
SOURCE: IMAGE CREDIT, FORD DESIGN.

△
**6.13** Early-stage high-performance Mustang design exploration
featuring widened track and extreme aero treatments
SOURCE: IMAGE CREDIT, FORD DESIGN.

△
**6.14** Early-stage high-performance Mustang design exploration illustrating progressive aerodynamic spoiler treatment, traditional stripes graphics, and rear quarter light insert panel
SOURCE: IMAGE CREDIT, FORD DESIGN.

The next few images show an early-stage data model that was completed in preparation to **mill** in **quarter scale** for executive review.

When designing a vehicle, the creative team will often explore derivatives and special versions to get an idea of the bandwidth or scope of a proposal's potential. This is important to evaluate a theme's ability to take on different applications in the future. A vehicle such as the Mustang needs to appeal to many different types of customers.

We all knew from the beginning of the design process we wanted the Mustang to be a worldlier car than any Mustang that came before. From a performance perspective it had to be the sort of machine that could easily take on the likes of Autobahn cruisers from premium German manufacturers

like BMW, Audi, and Mercedes. At the same time, it had to be able to compete with any fast and furious competition that was coming out of Asia. However, the diehard American **muscle car** fanatics could not be ignored. Although the purist with a good understanding of the history of the car would tell you that there was a significant difference between a typical muscle car and a pony car. What this amounted to was the vehicle couldn't be overly refined to lose a good deal of its wild horse, fun to drive, character. For motor sports enthusiasts who happened to also be the designers, this aspect of the project was more than a bit of fun (see Figures 6.13 and 6.14)!

Most car designers are automotive enthusiasts at heart, and in turn most enthusiasts love racing and racecars. Part of the magic of the Mustang is that it offers even

the least interested the thrill of motorsports and competition, if only from accelerating from a green traffic light! We were aware that the basic themes we were creating had to appeal to a wide variety of people, not just automotive enthusiasts. After all, this is the same sheet metal that had to look good at home or on the Nürburgring or parked at a Hertz rental car lot. In fact, we even joked that working on the project in general was essentially designing the world's coolest rental car! To stay fresh and enthusiastic as a designer, you always have to look for ways to have fun while doing your work. Imagining the next generation Mustang as a racecar definitely provided ample opportunity to accomplish that (see Figures 6.15 and 6.16).

△△
**6.15** Early-stage high-performance Mustang design
exploration featuring traditional auxiliary lamp
inserts in grill graphic
SOURCE: IMAGE CREDIT, FORD DESIGN.

△
**6.16** High-performance Mustang Design Study,
illustrating progressive lower aero treatment and
powerful hood bulge complete with upper air intakes
SOURCE: IMAGE CREDIT, FORD DESIGN.

## The Major League: Design Development in Full-Size

A key milestone in every project is the point the decision is made to go into full-size one-to-one **clay model** development. For a designer it also represents a significant victory. The management has decided that the proposal and ideas warrant further investigation. A team of clay **modelers** is assigned. **Engineers** begin study, verification, and evaluation of the concept. Studio resources are committed. Seemingly all at once there's lots of attention placed on a designer. It's a huge endorsement but it's also a friendly welcome to the hot seat. It suddenly goes from being a personal/individual investigation to a corporate undertaking (see Figures 6.17 and 6.18).

Along with sketching, many designers at Ford prefer to work in virtual **3-D** quite early on in the process. This gives us the ability to understand the relationship of the sculpture and theme of a vehicle to the engineering requirements. Once a decision is made, having those digital files allows the team to move quite quickly to mill the digital data in a clay model. Ultimately, however, we as designers do not stand next to a sketch or a data model. The final output is a tangible object that can be experienced in real life. There is no substitute for sculpting, taking the model outdoors, evaluating it and interrogating its qualities to every last detail. This process happens again and again, developing refinements and improvements, incorporating information from engineering. Reviewing it with your peers, taking it outside again and again, evolving, refining, refining! This in fact IS the core of the car design endeavor (see Figures 6.19 and 6.20).

The creative chemistry of a car design studio ebbs and flows between collaboration and competition. There were dozens of initial proposals for the Mustang. Fewer **scale models** were achieved. And at this point the design group was down to three remaining full-size proposals. These three proposals would be submitted to a **market research** event where the corporation would gain feedback on the acceptance of the design and guide next steps in the design development process. It is meant as an exercise to inform our intuition and balance our gut reactions with scientific data and research on what customers see in the design (see Figure 6.21).

In addition to the customer research clinics that the organization coordinates, there's a good deal of internal debating and discussion that goes on regarding the various theme proposals by

designers and senior management outside of design. As you can imagine with a car as important as the next generation Mustang, there are a lot of opinions to weigh and a designer must be ready to hear them all. Intuition and passion are at the core of this discussion. And as most of us are automotive enthusiasts, we've got the customers' wishes in mind. After all, everyone working on the program actually wanted to buy the car. The corporation prioritizes the satisfaction of the user and considers dealers an integral part of marketing a vehicle. As a result, the dealer council is also asked to weigh in on forthcoming products and their collective opinion is valued. This is the case in almost every major automotive manufacturer. Dealers are in fact on the front line of the marketplace and do have a unique feel for customer tastes, wants, and needs. One of the talking points that rose to the front of this discussion was the fine line and balancing act of whether the future Mustang was to be evolutionary or revolutionary.

◁◁◁
**6.19** Jordan Meadows taping rear tail lamp graphics on clay model proposal, Mustang Design Study
SOURCE: IMAGE CREDIT, FORD DESIGN.

◁◁
**6.20** Mustang Design Study, clay model development work in progress, complete with tape lines and graphics indicating alterations and changes
SOURCE: IMAGE CREDIT, FORD DESIGN.

◁
**6.21** Mustang Design Study, clay model prepared with dynoc treatment for management review
SOURCE: IMAGE CREDIT, FORD DESIGN.

Ultimately, design came back with the thought that in the case of premium brands like Porsche, for example, there is clearly an evolutionary approach. With so much history supporting this point of view, the theory was to remain true to one self and be the best version of that, rather than chasing trends and playing the competitors' game. Added to this was the fact that the 2015 Mustang had significantly better proportions than the existing car: lower, wider, etc. The longer hood and sportier stance were remarkably more athletic and efficient-looking than both the car it would replace and those of external competitors. In addition to this, the car would outperform the existing Mustang by every measure technically. The new car by our hypothesis would also conquer new customers from import manufacturers as well, due to its evolved, more premium, look and feel. As the discussion continued, these points were made evident in the design through sleepless nights, sweat and tears, tape, sketches, so on and so forth, as the refinement process continued (see Figures 6.22–6.24).

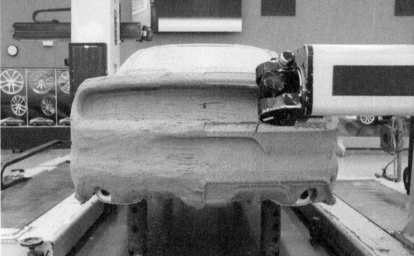

▷▷▷
**6.22** Mustang Design Study, clay model study prepared with dynoc, undergoing proportional alterations and changes indicated with tape
SOURCE: IMAGE CREDIT, FORD DESIGN.

▷▷
**6.23** Mustang Design Study, rear end study undergoing clay model milling operation
SOURCE: IMAGE CREDIT, FORD DESIGN.

▷
**6.24** Mustang Design Study, one-to-one clay model, receiving an updated milling and detail
SOURCE: IMAGE CREDIT, FORD DESIGN.

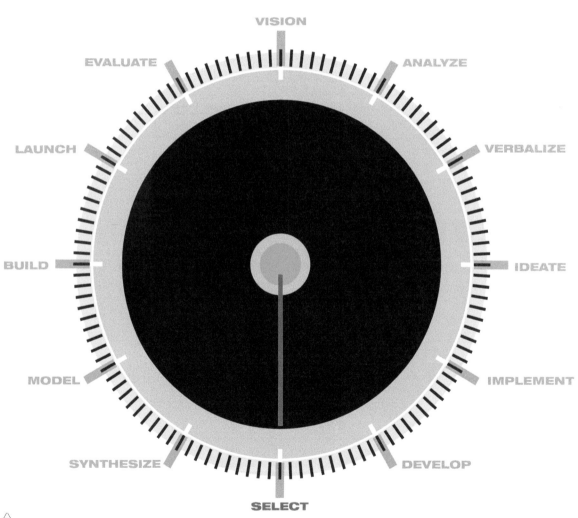

△
**7.0** Process locator gauge
SOURCE: IMAGE CREDIT, JORDAN MEADOWS.

# CHAPTER 7

**SELECT**
REFINING PROPOSALS: FORD MUSTANG CASE STUDY, PART 2

# SELECT

## Refining Proposals and Making a Final Selection: Ford Mustang Case Study, Part 2

**Refining Proposals and Making a Final Selection**

This chapter will explore how an organization makes a final choice on a product to move forward into production. In a sense this is the culmination of all of the various processes and efforts and development that have gone before it. This is the point when a team of decision-makers must consider the marketplace opportunity, target customers, and carefully decide which one of the proposals will best suit their wants and needs.

Different organizations have various methods for arriving at a selection. However, the brand, customer wants and needs, technical assumptions, and design goals all weigh heavily in the decision process. For the sake of continuity, we will continue with the Mustang as a case study of how an organization goes through a selection process (see Figure 7.0).

# CASE STUDY

||||||||||||||||||||||||||||||||||||||

# Ford Mustang
# Part 2

△
**7.1** Mustang Design Study, front three-quarter comparison of final proposals leading toward final selection
SOURCE: IMAGE CREDIT, FORD DESIGN.

Chapter 6 was dedicated to a case study on the evolution and development of a design proposal for the 2015 Ford Mustang. After a good deal of work, much of which parallels the processes outlined in this book, the corporation was down to three final proposals (see Figures 7.1 and 7.2). The design of these three proposals was vetted from a technical standpoint. Through an iterative process of analog **clay modeling**, **digital milling**, and graphic enhancements and developments, the design office had arrived at three proposals that everyone felt quite comfortable with. Because the Mustang has a good degree of history, and is composed of signature cues like the three-slot tail lamps, side scoops, and fastback roofline, much of the design elements were set in place. The corporation had also made a key decision to take the front-end design in a particular way that related the car to its overall family of products. In this regard it was almost like participating in a cooking competition where three chefs were given the same

ingredients and challenged to make similar dishes but executed in different ways. Millimeters mattered. They each evoked distinct moods and feelings based on the execution of their design. The time had come in the process to select one to concentrate on as this would be the single proposal released to engineering for production.

Ford Motor Company uses a combination of many different factors to select a theme to go into production. They include qualitative and quantitative market research and recommendations from various internal organizations such as Design, Engineering, Manufacturing, and Marketing. Senior leadership intuition and expertise are also combined with these inputs to make a decision. The process is lengthy and involves due diligence around multiple factors, technical rigor, and exploration of different possibilities. From this point forward a massive amount of coordination and funding is needed to put a vehicle on the

△
**7.2** Mustang Design Study, rear three-quarter comparison of final proposals leading toward final selection
SOURCE: IMAGE CREDIT, FORD DESIGN.

TAIL LAMPS ON

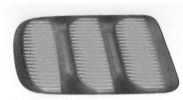

BRAKE LAMPS ON

road for the purchasing public, and much care is taken to get that decision right.

## Creating an Initial Design Prototype

Once the design is frozen, a preproduction model must be prepared for theme selection and evaluation. This is commonly referred to as a **hard model** or prototype, since they are most often constructed in fiberglass or resin or some sort of durable material that can be transported while maintaining dimensional fidelity. Every aspect of the model is meant to communicate the intention for production. It's often said that the devil lies in the details and in this case it's very true (see Figures 7.3–7.9). A successful automobile is a collection of hundreds of different parts, each with their own design content. Managing each one requires skill, talent, and expertise in the coordination of a team working together very well.

While most of the time and energy is going into executing the hard

SEQUENTIAL TURN SIGNALS

BACK UP LAMPS ON

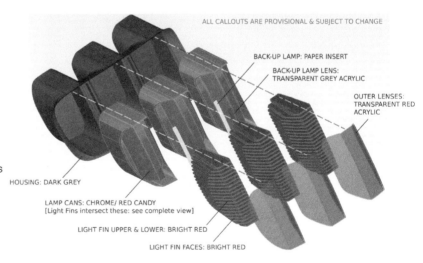

ALL CALLOUTS ARE PROVISIONAL & SUBJECT TO CHANGE

BACK-UP LAMP: PAPER INSERT

BACK-UP LAMP LENS: TRANSPARENT GREY ACRYLIC

OUTER LENSES: TRANSPARENT RED ACRYLIC

HOUSING: DARK GREY

LAMP CANS: CHROME/ RED CANDY
[Light Fins intersect these: see complete view]

LIGHT FIN UPPER & LOWER: BRIGHT RED

LIGHT FIN FACES: BRIGHT RED

△△
**7.3** Mustang Design Study, illustration of the signature sequential tail lamp feature
SOURCE: IMAGE CREDIT, FORD DESIGN.

△
**7.4** Mustang Design Study, exploded view illustration of tail light construction for design prototype
SOURCE: IMAGE CREDIT, FORD DESIGN.

◁◁◁ **7.5** Mustang Design Study, work-in-progress prototype grill insert featuring pony badge integration
SOURCE: IMAGE CREDIT, FORD DESIGN.

◁◁ **7.6** Mustang Design Study, prototype hard model being prepared for surface milling
SOURCE: IMAGE CREDIT, FORD DESIGN.

◁ **7.7** Mustang Design Study, prototype grill inserts featuring possible lamp integration
SOURCE: IMAGE CREDIT, FORD DESIGN.

▽▽ **7.8** Mustang Design Study, prototype hard model foam armature in process of being milled
SOURCE: IMAGE CREDIT, FORD DESIGN.

▽ **7.9** Mustang Design Study, prototype being prepared for final painting and assembly
SOURCE: IMAGE CREDIT, FORD DESIGN.

model for evaluation and selection, there is also a significant amount of work in pursuing variations of the theme. A successful car is the result of many individual subordinate parts all harmoniously blended together. Changing one of these can alter the impression of the car altogether. One of the key features that was debated and discussed involved a single continuous line without a haunch running from start to finish of the vehicle. This gave a clean, concise impression to the body side; in a sense it also hearkened back to the 2005 car. The following series of images explored what that composition would look like with a more traditional haunch over the rear fender (see Figures 7.10 and 7.11).

### Final Theme Selection

After the Mustang development team had conducted extensive qualitative, quantitative consumer and market research, a decision was made to pursue a direction. However, it was quite clear that further design development needed to take place. In Ford corporate terms, we had arrived at a "Go for One" decision. The team would be recomposed and efforts would now be focused on a single proposal (see Figures 7.12–7.21).

After the research event, the designers were asked to do a blend of the best attributes of all the submitted proposals. The following series of images were shown to management as a proposal for a blend of the various elements that would make for a fourth theme to go forward into production (Figure 7.22). Of note, the front-end development was the most aggressive version of what we attempted but without the running lamps that were

seen as hearkening back to the outgoing car and classical. The composition would keep the body side and aero-efficient approach to the roofline from some of the initial proposals but adding a rear haunch. Finally, a graphic coast-to-coast Mustang element would finish off the rear and reference classic Mustangs of the past; this also helped to reinforce a Mustang impression as we had chosen to edit the cues elsewhere, such as the hockey-stick element on the side and signature rear quarter vents in the **B-pillar** area. Through a series of ongoing discussions,

△△
**7.10** Mustang Design Study, illustrated without rear fender haunch treatment
SOURCE: IMAGE CREDIT, FORD DESIGN.

△
**7.11** Mustang Design Study, explored with rear fender haunch treatment
SOURCE: IMAGE CREDIT, FORD DESIGN.

**SELECT**
REFINING PROPOSALS: FORD MUSTANG CASE STUDY, PART 2

◁◁
**7.12** Mustang Design Study, front three-quarter image of one of the final themes considered for production, highlighting forward-swept headlamps and vertical side vents
SOURCE: IMAGE CREDIT, FORD DESIGN.

◁
**7.13** Mustang Design Study, rear three-quarter image of one of the final themes considered for production, highlighting an aeronautical roofline and body color treatment between rear tail lamps
SOURCE: IMAGE CREDIT, FORD DESIGN.

△
**7.14** Mustang Design Study, side view comparison of one of the final themes considered for production, highlighting the development and evolution versus the outgoing car
SOURCE: IMAGE CREDIT, FORD DESIGN.

△△△
**7.15** Mustang Design Study, rear three-quarter image of one of the final themes considered for production, highlighting a tall proportion to the signature three-bar tail lamps
SOURCE: IMAGE CREDIT, FORD DESIGN.

△△
**7.16** Mustang Design Study, front three-quarter image depicting one of the final themes considered for production, highlighting rear-swept headlamps and full hood sections
SOURCE: IMAGE CREDIT, FORD DESIGN.

△
**7.17** Mustang Design Study, side view comparison of one of the final themes considered for production, highlighting the development and evolution versus the outgoing car
SOURCE: IMAGE CREDIT, FORD DESIGN.

△△△
**7.18** Mustang Design Study, front three-quarter image of one of the final themes considered for production, highlighting angular sections and unified lower graphics
SOURCE: IMAGE CREDIT, FORD DESIGN.

△△
**7.19** Mustang Design Study, rear three-quarter image depicting one of the final themes considered for production, featuring a horizontal black graphic unifying the signature three-bar tail lamps
SOURCE: IMAGE CREDIT, FORD DESIGN.

△
**7.20** Mustang Design Study, side view comparison depicting one of the final themes considered for production, highlighting the development and evolution versus the outgoing car
SOURCE: IMAGE CREDIT, FORD DESIGN.

Jordan Meadows

△
**7.22** Mustang Design Study, rear three-quarter top view of post-selection theme, blending key attributes combined to create a unified proposal
SOURCE: IMAGE CREDIT, FORD DESIGN.

the engineering group developed more proportional modifications to make the layout and architecture of the Mustang even stronger. The track was increased to allow for more sculpture in the body side. This also gave us the room to create more drama, sculpture, and visual impact by the addition of very generous fenders.

On the design side, the team rallied around the selected theme, contributing to a collective proposal that combined input from the best aspects of proposals up to that point in the process. The evolution and distilling of

◁
**7.21** Mustang Design Study, side view comparisons of the final three themes which were considered for production
SOURCE: IMAGE CREDIT, FORD DESIGN.

the various inputs continued as the group moved forward. For example, many on the team loved the coast-to-coast graphic that encapsulated the three-bar elements in an aeronautical fashion. We likened them to afterburners on a jet. In the rear quarter we arrived at a solution that would allow personalization and altering the center panels for color versus all glass. Most of the themes that were executed had a rear haunch over the rear axle. Ultimately the decision was made to go with it in the final production model. Also, we decided to go with a contemporary approach to the sill feature rather than having a traditional Mustang hockey stick. The final blend was quickly taking shape. Senior management requested that we should strongly indicate a change

from the outgoing car to appease the concerns of the car appearing too similar or too evolutionary. The final daylight opening (DLO) was an edit done with this in mind. This would eliminate the traditional Mustang B-pillar or vent treatment (Figure 7.23). A decision was made to also do without the round rally style lamps located in the main grill opening as this was also seen to be a carryover element from the outgoing car. Finally, to complete the composition, we went with the front end proportion and execution that mostly related to the Dearborn theme but integrated developments and enhancements from the other designers in the lower fascia.

△
**7.23** Mustang Design Study, side view of post-selection theme,
blending key attributes combined to create a unified proposal
SOURCE: IMAGE CREDIT, FORD DESIGN.

*Jordan Meadows*

◁

**7.24** The final production Mustang released to the public seamlessly blending the hard work and passion of a group of dedicated professionals
SOURCE: IMAGE CREDIT, FORD DESIGN.

△
**7.25** Rear detail view of signature three-bar tail lamps from final production Mustang
SOURCE: IMAGE CREDIT, FORD DESIGN.

## The Final Cut

After receiving various inputs from the design team and blending the best attributes, the Mustang design leadership had arrived at a final composition. This point in the process is known as a design freeze. This also marks the point that the selected theme is released to the engineering group for the process of feasibility and production execution. This is the stage of the game where collaboration between the creative teams and the engineering teams is most crucial. Again, the devil is in the details—with regard to final design execution this is very much the case! While delivering the sensational-looking car that was expected, the team was also well aware that the Mustang had to have fantastic craftsmanship. It needed to communicate the progressive technology that it offered. Much of this was done with the detailing component design: elements like the lights on the front and rear, for example. They retain the traditional three-bar graphic element. This was a traditional Mustang cue that the design team felt strongly about keeping. However, they use new materials, textures, and finishes to convey a fresh, modern feel.

The design team continues to evolve and develop clay models that represent the production intent. A team of **engineers**, marketing people, and designers and manufacturing representatives are constantly evaluating these clay models, monitoring minute details, ensuring that these changes can be processed and manufactured. These clay models are then used to develop a second round of highly detailed hard models and prototypes. These models are then sent for final review with senior management, dealers, and are also evaluated by customers in research clinics for market acceptance. The process of designing a car is a lengthy one that can take years. These ongoing evaluations ensure that the group is on track to deliver what they set out to do at the beginning of the program.

With some projects where there is a good deal of passion, the team can actually over-achieve. Such was the case with the final rendition of the Mustang that was prepared for production. The feedback to these final reviews was exactly what was hoped for by the team. They had created a fast, fun, attainable car with a sensational appearance. This was a result of great collaboration by talented enthusiasts.

The Mustang is not only a car but also a cultural artifact, a true piece of American pop culture. So in that spirit we revealed the car at several high profile locations around the globe such as Times Square in New York, and TCL's Chinese Theatre in Hollywood. I can only hope to contribute to a car as significant as this again, but it's safe to say I probably won't see a design I helped with on that famous red carpet. It was truly a once-in-a-lifetime experience that I'm grateful to have had the opportunity to enjoy (see Figures 7.24–7.27).

**Acknowledgments**
Special thanks to: Raj Nair, Derek Kuzak, J. Mays, Moray Callum, Freeman Thomas, Joel Piaskowski, Darrell Beamer, Tim Boyd, Greg Hutting, Shelby Faulhaber, John Clinard, Francesca Montini, Craig von Essen, Jennifer Flake, and the entire PD and Communications teams at Ford Motor Company!

△ △
**7.26** Front three-quarter view of production Mustang
SOURCE: IMAGE CREDIT, FORD DESIGN.

△
**7.27** Rear three-quarter view of production Mustang
SOURCE: IMAGE CREDIT, FORD DESIGN.

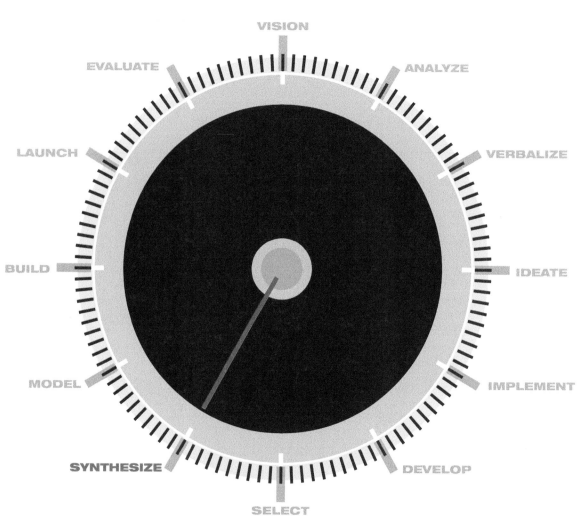

△
**8.0** Process locator gauge
SOURCE: IMAGE CREDIT, JORDAN MEADOWS.

**SYNTHESIZE**
BRINGING IDEAS TOGETHER

# CHAPTER 8

# SYNTHESIZE

Bringing Ideas Together, Component Design,
and the Application of UX

**Bringing Ideas Together, Creating an Overall Composition**

In the previous chapters of this book we have discussed the evolution and selection of a key theme for a vehicle. Deep diving into the 2015 Mustang as an example, we followed how a product development team came to select an exterior design for production. Theme selection represents a point in the creative journey where all the conceptual development that came before is given a level of solidity. However, after a lengthy process involving a broad team, the group is committed to only one, albeit very important, aspect of the vehicle's design.

Vehicles are incredibly complex by nature. And they involve the development of multiple simultaneous work streams. With regard to the Mustang, for example, a similarly intense effort was conducted for the interior of the vehicle in parallel with the exterior. And similarly, there were very comprehensive efforts dedicated to performance and engineering aspects of the vehicle. All of these would combine to deliver a unique and meaningful experience for the Mustang user. Be it trains, planes, motorcycles, or boats, all forms of vehicle design require that teams seamlessly blend multiple work streams to create one holistic product (see Figure 8.0).

This chapter is dedicated to the art and science of blending those different disciplines. It represents a point in the journey where the designers must become intensely focused on delivering an impactful **UX (User Experience)**. UX design is a term common in almost all forms of commercial design and in some regards is quite broad. The field is fast-moving and involves an appreciation of all the constituent parts required for a meaningful encounter with a product. In fact, one could spend an entire career dedicated specifically to UX design. Though it is placed here in this book to show how different work streams can be merged, the actual UX process began with a thorough user analysis that we discussed in Chapter 2. However, the focus of this chapter is to cover the field in general overview, relating specifically to vehicles. And while the term itself is quite broad, it has specific applications to **aesthetic principles in transportation design**.

△
**8.1** Jeep Willys concept, featuring an all-weather interior seating concept eliminating traditional upholstery and trim
SOURCE: IMAGE CREDIT, FCA/JEEP DESIGN.

Vehicles are an assemblage of many different products. They require a symphony of creative efforts. Each part must be designed individually with merits on its own, yet seamlessly and harmoniously integrate with the other parts to create a cohesive statement. This requires special strategies and techniques for blending coordinated elements to form a vehicle that delivers its intended experience.

## UX (User Experience) Design in Overview

What is UX design and how does it specifically relate to vehicles? Due to the wide usage of the term and the fact that it encompasses many things potentially, it's quite easy to be confused and get lost in the subject. However, the true focus is quite simple. UX design is the practice of enhancing user satisfaction by generating and developing systems to add value to a user's interaction with the vehicle.

The area of focus came as an outgrowth of the study of human factors and ergonomics in industrial and product design. During the latter half of the twentieth century as products became more technologically advanced, the field evolved specifically with regard to **human–machine interfaces (HMI)**. The proliferation of consumer-electronics then mandated a specific level of focus on graphical user interfaces (GUI) and how information is communicated

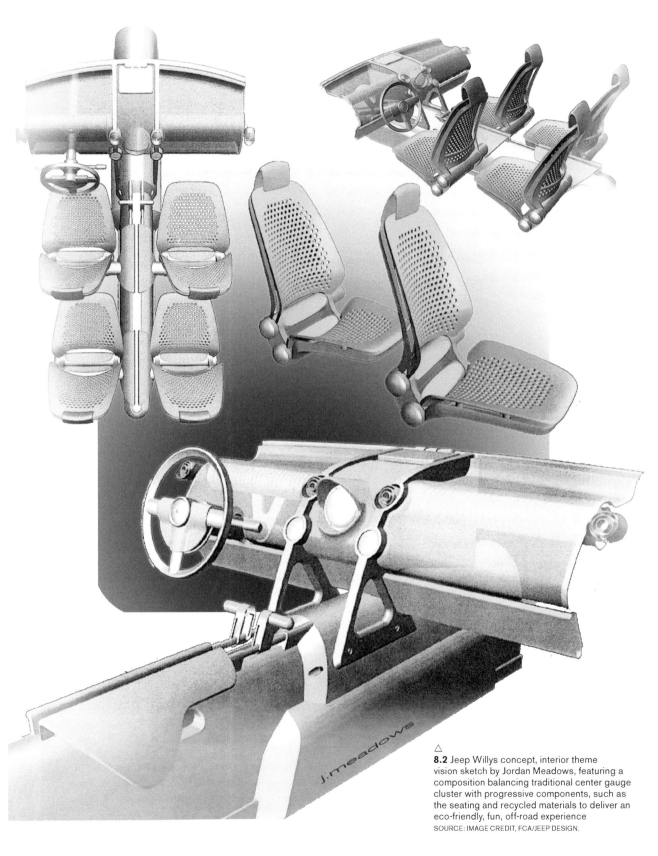

△
**8.2** Jeep Willys concept, interior theme vision sketch by Jordan Meadows, featuring a composition balancing traditional center gauge cluster with progressive components, such as the seating and recycled materials to deliver an eco-friendly, fun, off-road experience
SOURCE: IMAGE CREDIT, FCA/JEEP DESIGN.

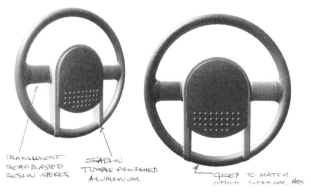

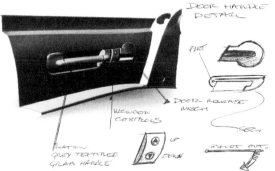

through GUI. Presently, due to the amount of design required for web-based and mobile applications, the field has grown to encompass these areas of focus as well. Though the roots of UX have been around for quite some time, the actual term was coined by Don Norman in the original edition of his book, *The Design of Everyday Things*.[1] It's a must read for any designer involved in a field where UX is crucial, as currently the term has been stretched to include nearly every aspect of how a user interacts with a product.

In terms of hardware, UX is most relevant and applicable to interior design with regard to vehicles. It has specific impact to the HMI within the vehicle (see Figures 8.1 and 8.2). By extension it also includes nearly every ergonomic equation and touch point. This goes from the thickness of the steering wheel and how many buttons are on it, to the types

of gauge clusters, the **HVAC** controls, and literally every aspect of the vehicle that a user might come in contact with (see Figures 8.3 and 8.4). UX covers not only what specifically is seen on every screen in a vehicle, it also involves information architecture and the graphic communication of every function the vehicle provides. One can then quickly see that with a product as complex as a vehicle, effectively designing the UX is crucial to its success.

## Experiences in Motion: UX for Vehicles

While UX design becomes highly impactful in interiors, it expands to all aspects of a vehicle's design and how it is used. This also includes the exterior. And since we understand now that vehicles are by nature a collection of many related products that have to work well together, exterior function must be informed by interiors and vice versa (see Figures

△
**8.3** Jeep Willys concept, excerpts from the build sketch book by Jordan Meadows, featuring steering wheel, door trim, seating, and instrument panel component callouts
SOURCE: IMAGE CREDIT, FCA/JEEP DESIGN.

▷▷
**8.4** Jeep Willys concept, excerpts from the build sketch book by Jordan Meadows, featuring rear lighting details, HVAC controls and retractable GPS navigation screen
SOURCE: IMAGE CREDIT, FCA/JEEP DESIGN.

▷
**8.5** Jeep Willys concept, interior seen here with removable seating covers
SOURCE: IMAGE CREDIT, FCA/JEEP DESIGN.

8.5 and 8.6). In some cases, such as motorcycles, there is no boundary at all. The exterior design of daylight openings that have an enormous impact on a graphic read of the outside of the vehicle will also impact the user experience that the driver and occupants have by seeing

VISOR COVER AND HINGES TO BE FINISHED SAME AS FRAME MEMBERS. SCREEN BOX TO BE SAME AS RUBBERIZED GREY CENTER STACK MATERIAL.

S. MEADOWS

FOLD AWAY INTERFACE SCREEN FOR G.P.S. AND ENTERTAINMENT. TO BE BACKLIT IN BRITE GREEN. SEE INDIGLO WATCH FOR DETAIL.

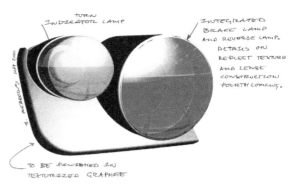

TURN INDICATOR LAMP

INTEGRATED BRAKE LAMP AND REVERSE LAMP. DETAILS ON REFLECT TEXTURE AND LENSE CONSTRUCTION FOURTH COMING.

TO BE FINISHED IN TEXTURIZED GRAPHITE

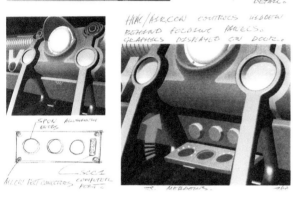

HVAC/AIRCON CONTROLS HIDDEN BEHIND FOLDING PANELS. GRAPHICS DISPLAYED ON DOOR.

SPUN ALUMINIUM KNOBS

SCC1 ACCESS PORT COMPUTER PORT.

S. MEADOWS.

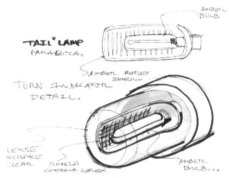

AMBER BULB

TAIL LAMP PARABOLA.

AMBER REFLEX SHIELD...

TURN INDICATOR DETAIL.

LENSE SURFACE CLEAR

SHIELD CONTAINS REFLEX

AMBER BULB...

△
**8.6** Jeep Willys concept, rear three-quarter top view, illustrating harmonious blend of interior and exterior designed for a dedicated user experience. The concept makes use of lightweight aluminum and recycled plastic materials that are communicated in its design semantics with consistency throughout the vehicle
SOURCE: IMAGE CREDIT, FCA/JEEP DESIGN.

out of it. Each subcomponent on the exterior offers an opportunity for a unique UX. An exterior door handle not only allows for entry into the machine, it also has a haptic quality. It's a crucial touch point that communicates a certain tactile message. Lighting features can bring a jewel-like sense of precision, detailed scale, and visual interest to the corner of a vehicle. They are also an

opportunity for pedestrians and other drivers to have a unique experience with that vehicle. For example, the sequential illumination on a Mustang's tail lamp as it indicates a turn is a memorable moment for any other driver viewing the car.

Along with shape and light, sound is an undeniably important aspect of delivering a memorable and

meaningful experience. Harley-Davidson takes this idea very seriously with regards to the sound of its motors. Conversely, Tesla trades on the fact that it delivers hardly any sound or vibration owing to its lack of a combustion **powertrain**. Both are not only crucial user experiences directly aligned with the product narrative, but also expressions of the brand. Entry and egress features and door systems are also crucial areas for UX. One of the key selling points of a Jeep is its removable roof, doors, and the functionality and enjoyment that provides. Doors not only provide a means for entering the vehicle, they are a threshold for UX. In fact, any transition point in a designed object is an opportunity to make a statement. When taken altogether, these various statements should ideally be interwoven and arranged to form a harmonious and satisfying expression of the product narrative and brand.

UX design for vehicles also poses unique challenges and concerns. Because vehicles are in motion and potentially dangerous, safety and regulatory issues are a top priority. Designers need to provide crucial information while not posing a hazardous distraction. The flow of information, its intended usage and interaction must be intuitive. Usability and ease of operation are of the utmost importance so drivers can access relevant data and info while maintaining safe command of the machine. To accomplish this, it's best to use a primary, secondary, and subordinate strategy for the delivery of key information. For example, vehicle speed should always be the largest and most legible bit of information communicated as close to the basic **occupant sightlines** while driving. Secondary information such as temperature and engine performance should be all delivered with less priority. Third, subordinate details such as entertainment can be smallest and furthest away from the driver's line of sight. Additionally, key controls should be durable, robust, and reliable. Where

possible, they should also provide haptic feedback and redundant indications of usage. For instance, gear shift and PRNDL levers that make a precise sound and have weight and limited resistance, are preferable. Crucial controls such as lighting and indicators should always have an analog switch that the driver can touch, feel, and acknowledge when in use.

In addition to this, UX design for vehicles needs to be seamless. The graphic aesthetic and family of shapes given to the forms should be cohesive with the overall look and feel of the vehicle. This communicates a sense of unity and craftsmanship important for eliciting trust and ease from the driver. The controls and data points should also be seamless in their ability to integrate external applications. Users require the ability to accommodate various mobile and external devices. Finally, all the technology-based aspects of UX for vehicles should be in some way upgradable. Vehicle development timeframes are typically much longer than consumer-electronic development timeframes. Users want and need the ability to mitigate the difference, updating tech-based elements without changing the entire vehicle.

**Strategies and Methods for Vehicle UX**

When embarking on the process of developing the UX design for any vehicle, it's best to re-familiarize yourself with the overall vehicle **design brief**. Coordination and harmony with the overall theme are the results of a good understanding of the established user persona and narrative. After all, every aspect of the vehicle should share

a common target. With this established, it's good to then take another look at the aesthetic objectives and consider how they relate specifically to the UX. Image boards and visual inspirations can be quite valuable at this point in the process. Typically, many choices on color, graphic composition, and even font/type choice can be informed by a good image board. Finally, get reacquainted with any **storyboarding** that has previously been executed. They were the early steps of this exercise. *EMPATHY and understanding how the user needs to FEEL will be crucial in developing your UX!*

Dive deeper into any pre-existing storyboarding to create a detailed journey map. This can be as basic as a list of all the functional objectives a user may want to achieve in chronological order. It's important to be as detailed as possible, deconstructing any and every desired task and function. This journey map can be verbal or visual but it should ideally be both. It should also begin with the user's intention, and conclude with how they achieve it. It is also crucial to note any point along the way that poses a challenge or obstacle. Eliminating these points of frustration is very much the goal at this stage of the game. This is when designers employ good old-fashioned problem-solving to enhance user satisfaction and create ideal outcomes. Necessity is the mother of invention. In generating a journey map, a designer should be seeking any point along the process where the user finds that necessities are not provided for (see Figure 8.7).

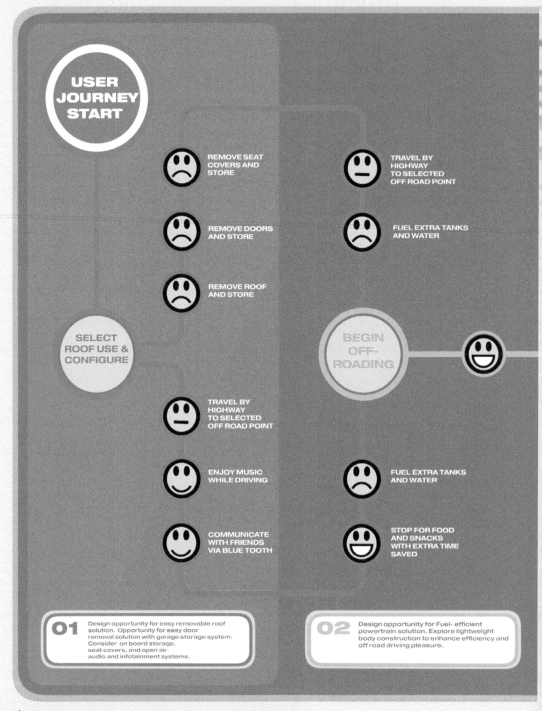

USER
JOURNEY
START

REMOVE SEAT
COVERS AND
STORE

TRAVEL BY
HIGHWAY
TO SELECTED
OFF ROAD POINT

REMOVE DOORS
AND STORE

FUEL EXTRA TANKS
AND WATER

REMOVE ROOF
AND STORE

SELECT
ROOF USE &
CONFIGURE

BEGIN
OFF-
ROADING

TRAVEL BY
HIGHWAY
TO SELECTED
OFF ROAD POINT

ENJOY MUSIC
WHILE DRIVING

FUEL EXTRA TANKS
AND WATER

COMMUNICATE
WITH FRIENDS
VIA BLUE TOOTH

STOP FOR FOOD
AND SNACKS
WITH EXTRA TIME
SAVED

**01** Design opportunity for easy removable roof
solution. Opportunity for easy door
removal solution with garage storage system.
Consider on board storage,
seat covers, and open air
audio and infotainment systems.

**02** Design opportunity for Fuel- efficient
powertrain solution. Explore lightweight
body construction to enhance efficiency and
off road driving pleasure.

△
**8.7** Example of the journey map containing the essential usage
channels, touch points, and emotional reactions. The lower section of
the chart outlines opportunities for design solutions. Journey mapping
should always be done with a clear understanding of an intended user.
SOURCE: IMAGE CREDIT, JORDAN MEADOWS.

**SYNTHESIZE**
BRINGING IDEAS TOGETHER

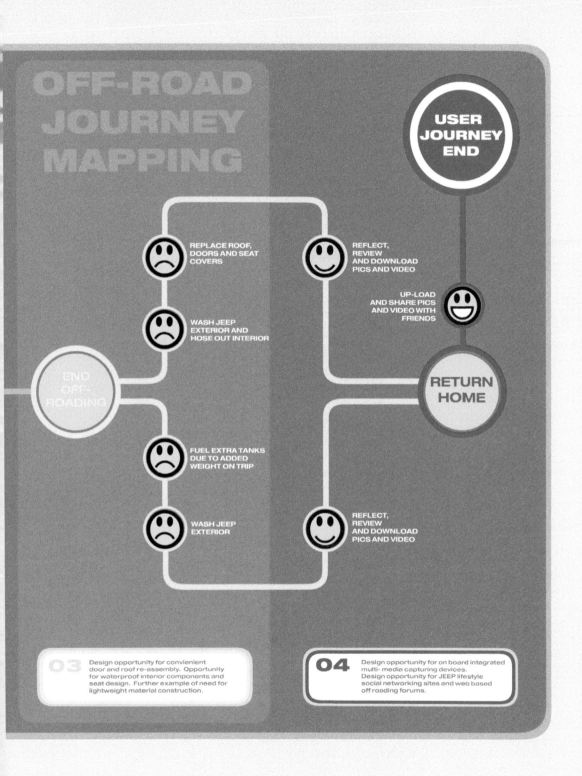

OFF-ROAD JOURNEY MAPPING

USER JOURNEY END

REPLACE ROOF, DOORS AND SEAT COVERS

REFLECT, REVIEW AND DOWNLOAD PICS AND VIDEO

UP-LOAD AND SHARE PICS AND VIDEO WITH FRIENDS

WASH JEEP EXTERIOR AND HOSE OUT INTERIOR

END OFF-ROADING

RETURN HOME

FUEL EXTRA TANKS DUE TO ADDED WEIGHT ON TRIP

WASH JEEP EXTERIOR

REFLECT, REVIEW AND DOWNLOAD PICS AND VIDEO

**03** Design opportunity for convienient door and roof re-assembly. Opportunity for waterproof interior components and seat design. Further example of need for lightweight material construction.

**04** Design opportunity for on board integrated multi- media capturing devices. Design opportunity for JEEP lifestyle social networking sites and web based off roading forums.

Once a detailed journey map has been created, one should then create an audit of existing assets. This will include any ergonomic and fixed hardware elements of the vehicle package already provided. It may also include any pre-existing graphical user interfaces, screens, cameras, speakers, and audio communication devices. Once an audit of existing assets has been concluded, any additional elements required to accomplish the intended task can be proposed.

With this in place, one can then begin to generate key framing and system mapping for the graphical user interfaces (GUI). This is effectively sketching out the interaction framework and high-level structure of screen layouts along with the various product flow, behavior, and organization. In essence, you're creating a sequence of key frame instructions to guide the user through an ideal sequence of events. For example, when the key fob button is touched, the interior lighting in the vehicle can gradually illuminate. The center console screen can then welcome the driver by name and ask what type of music they would like. Then when the key is inserted into the ignition, or the start button has been pressed, the lighting can slowly decrease while the vehicle exterior running lights can automatically turn on. Then the infotainment system begins to automatically play the driver's preferred music selection.

Once the wireframe order of operations and all of the necessary assets have been coordinated to create an ideal flow, critique the UX sharing with one's peers to hone and develop every aspect so it is user-friendly and seamless.

Once prototypes or mockups have been created, one can then conduct simulations to fine-tune the experience and flow. Finally, continue to evaluate and make sure that the solutions are aligned with the initial goal and product narrative.

### Looking Forward: Unique Opportunities

One of the things that makes vehicles both special and exciting to design is that they offer the opportunity to use multiple products coordinated together to provide an intended outcome. In this regard it's similar to music in that a conductor organizes different instrument sections to arrange a composition. UX design is very similar in that the designer is ultimately using all the available components of a vehicle to deliver user satisfaction. It is very much a craft of harmonious integration. The design brief and product narrative inform the process at every step of the way. As vehicles become more complex and integrate new and emergent technologies, the UX will become ever more important to their success, effectively adding an exciting layer of interaction to the aesthetic principles of transportation design. However, good UX design can also be low tech and very much about simple back-to-basics fun. Figures 8.8–8.20 are an example of synthesized elements forming the components of a Jeep Willys Concept: Component Design Case Study experience. They include excerpts from a build sketch book which is an essential part of completing the componentry design in a vehicle, and also synthesizing all of its varied parts.

### Note

1 Norman, Don. *The Design of Everyday Things* (New York: Basic Books, 2002).

||||||||||||||||||||||||||||||||||||||

# Jeep Willys Concept
## Component Design

△
**8.8** Jeep Willys concept, exterior front view featuring its unmistakable DNA
SOURCE: IMAGE CREDIT, FCA/JEEP DESIGN.

Provided on the coming pages are excerpts from the build sketch book by Jordan Meadows for the Jeep Willys concept (see Figures 8.8–8.20).

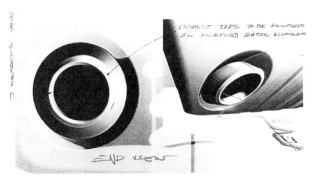

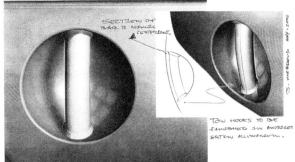

EXHAUST TIPS TO BE FINISHED IN ANODIZED SATIN ALUMINUM

END VIEW

SECTION OF BAR TO NEGATE CENTERLINE.

TOW HOOKS TO BE FINISHED IN ANODIZED SATIN ALUMINUM.

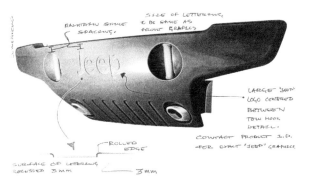

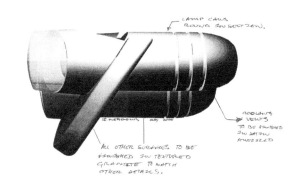

MAINTAIN SAME SPACING.

SIZE OF LETTERING TO BE SAME AS FRONT GRAPHICS

LARGE "JEEP" LOGO CENTERED BETWEEN TOW HOOK DETAIL.

CONTACT PRODUCT I.D. FOR EXACT "JEEP" GRAPHICS

ROLLED EDGE

SURFACE OF LETTERING RECESSED 3MM

3MM

LAMP CANS ROUND IN SECTION.

COOLANT VENTS TO BE FINISHED IN SATIN ANODIZED

ALL OTHER SURFACES TO BE FINISHED IN TEXTURED GRAPHITE TO MATCH OTHER DETAILS.

◁◁
**8.9** Jeep Willys concept, exterior theme sketches by Jordan Meadows, designed for the young and the young at heart, the exterior composition is meant to communicate core Jeep values delivered in a progressive way, providing the user with a means to experience off-roading in an eco-friendly way
SOURCE: IMAGE CREDIT, FCA/JEEP DESIGN.

◁
**8.10** Jeep Willys concept, exterior front three-quarter view, seen here fitted with hard top and multifunctional roof rack system
SOURCE: IMAGE CREDIT, FCA/JEEP DESIGN.

△△
**8.11** Jeep Willys concept, excerpts from the build sketch book by Jordan Meadows, featuring rear lighting tow hook and lower fascia exhaust details
SOURCE: IMAGE CREDIT, FCA/JEEP DESIGN.

△
**8.12** Jeep Willys concept, exterior rear three-quarter upper view, seen here fitted with hard top and multifunctional roof rack system
SOURCE: IMAGE CREDIT, FCA/JEEP DESIGN.

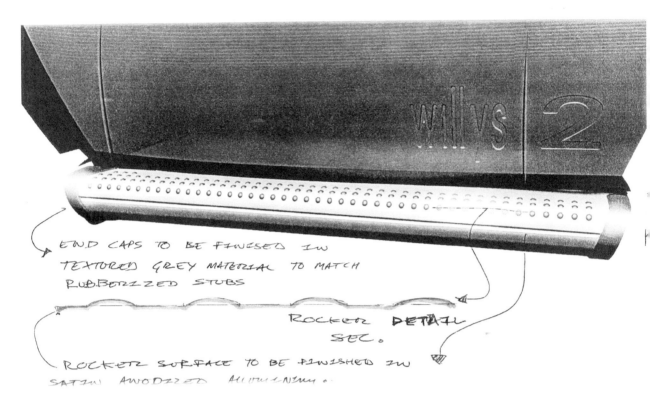

END CAPS TO BE FINISHED IN
TEXTURED GREY MATERIAL TO MATCH
RUBBERIZED STUBS

ROCKER DETAIL
SEC.

ROCKER SURFACE TO BE FINISHED IN
SATIN ANODIZED ALUMINIUM.

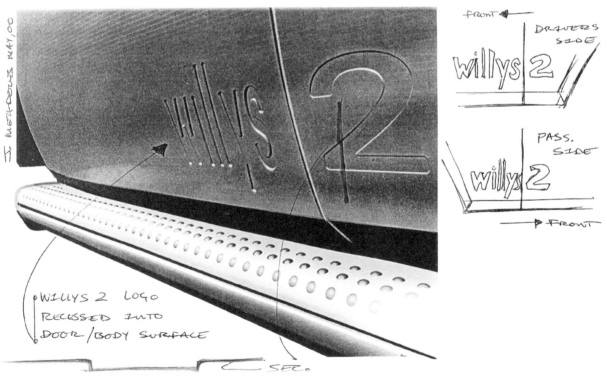

H. MCHRDOUS MAY, OO

WILLYS 2 LOGO
RECESSED INTO
DOOR/BODY SURFACE

SEC.

FRONT

DRIVERS SIDE

willys 2

PASS. SIDE

willys 2

FRONT

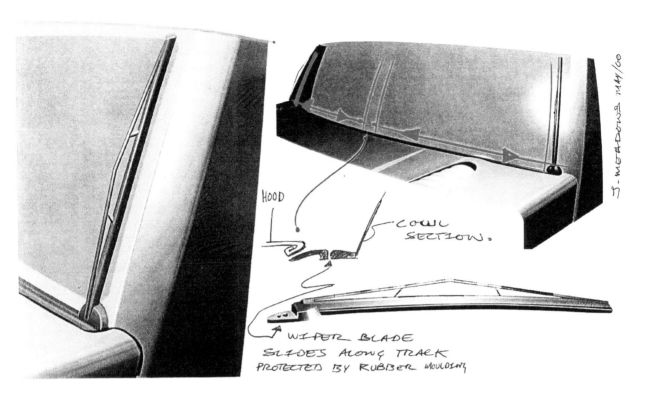

HOOD

COWL
SECTION.

J. MEADOWS MAY/00

WIPER BLADE
SLIDES ALONG TRACK
PROTECTED BY RUBBER MOULDING

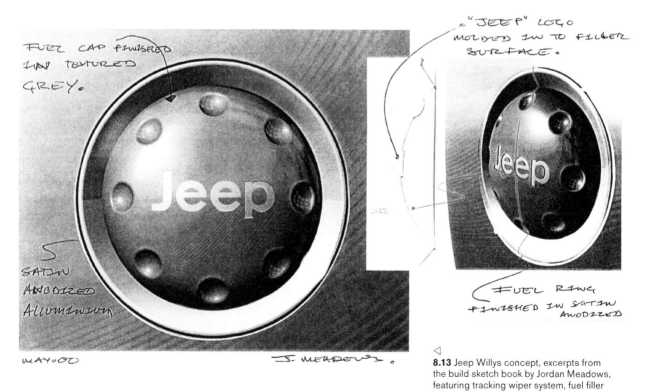

FUEL CAP FINISHED
IN TEXTURED
GREY.

"JEEP" LOGO
MOLDED INTO FILLER
SURFACE.

SATIN
ANODIZED
ALLUMINIUM

FUEL RING
FINISHED IN SATIN
ANODIZED

MAY/00

J. MEADOWS.

◁

**8.13** Jeep Willys concept, excerpts from
the build sketch book by Jordan Meadows,
featuring tracking wiper system, fuel filler
cap, and rocker step detail
SOURCE: IMAGE CREDIT, FCA/JEEP DESIGN.

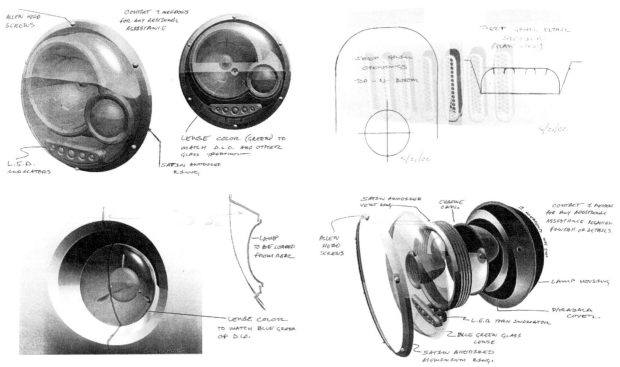

△
**8.14** Jeep Willys concept, excerpts from the build sketch book by Jordan Meadows, featuring exploded view of headlight details, and grill inserts
SOURCE: IMAGE CREDIT, FCA/JEEP DESIGN.

▷▷
**8.15** Jeep Willys concept, excerpts from the build sketch book by Jordan Meadows, featuring antenna, lighting, and grill mesh inserts
SOURCE: IMAGE CREDIT, FCA/JEEP DESIGN.

▷
**8.16** Jeep Willys concept, excerpts from the build sketch book by Jordan Meadows, featuring perforated header visor concept. This component allows for angle visibility to see traffic lights, etc. while maintaining an upright windshield beneficial for off-road. The perforations also reinforce and communicate the idea of lightness expressed with consistency on every aspect of the vehicle.
SOURCE: IMAGE CREDIT, FCA/JEEP DESIGN.

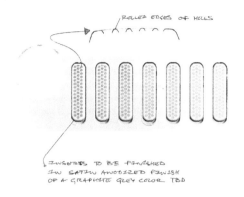

ROLLED EDGES OF HOLES

INSERTS TO BE FINISHED
IN SATIN ANODIZED FINISH
OF A GRAPHITE GREY COLOR TBD

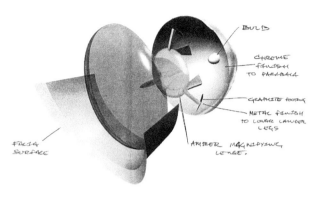

BULB

CHROME
FINISH
TO PARTBALL

GRAPHITE FOOTING

METAL FINISH
TO LOWER LANDER
LEGS

AMBER MAGNIFYING
LENSE

FASCIA
SURFACE

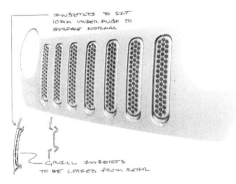

INSERTS TO SIT
10MM UNDER FLUSH TO
SURFACE NORMAL

GRILL INSERTS
TO BE LOADED FROM REAR

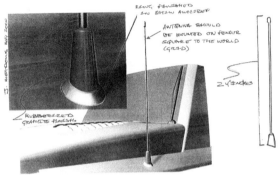

RING, FINISHED
IN SATIN ANODIZED

ANTENNA SHOULD
BE MOUNTED ON FENDER
SQUARE TO THE WORLD
(GRID)

RUBBERIZED
GRAPHITE FINISH

2.4" INCHES

PERFORATIONS
SHOULD BE MOLDED
INTO ROOF SURFACE
AND IMMULATE
FURNITURE

ALL EDGES OF
PERFORATIONS SHOULD
BE ROLLED.
CONTACT I. MEADOWS
FOR EXACT SIZE AND
SPACING OF HOLES.

CAPER

PERFORATIONS

ROOF SEC.

CLEAR PLEXI COLORED
TO MATCH TONE OF
WINDSHIELD.

WINDSHIELD.

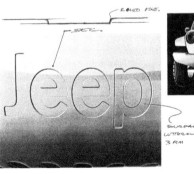

ROLLED EDGE.

SEC.

SURFACE OF
LETTERING, RECESSED.
3 MM

LARGE JEEP
LOGO CENTERED
BETWEEN
FOR LAMPS.
CONTACT PRODUCT
I.D. FOR
EXACT "JEEP" GRAPHIC

WITH TOP ON
PLEXI / AND GLASS SHOULD
MATCH IN COLOR AND
READ AS ONE SURFACE.

ENTIRE ROOF SHOULD
BE PAINTED EXTERIOR
BODY COLOR WITHOUT
CUT LINES OR CHANGE
IN TEXTURE.

CONTACT I. MEADOWS
FOR HELP, AND ADDITIONAL
DETAILS REGARDING
NUMBER AND SPACING
OF HOLES.

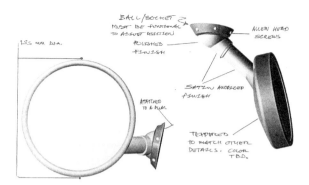

155 MM DIA.

BALL/SOCKET MUST BE FUNCTIONAL TO ADJUST POSITION. POLISHED FINISH

ALLEN HEAD SCREWS

ATTACHED TO A PILLAR

SATIN ANODIZED FINISH

TEXTURED TO MATCH OTHER DETAILS. COLOR T.B.D.

VENTS TO BE FINISHED IN SATIN ANODIZED ALUMINUM.

CONTACT J. MEADOWS ON EXACT DETAILS OF SHAPE OF OPENINGS FOR VENTS

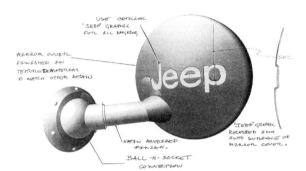

USE OFFICIAL "JEEP" GRAPHIC FOR ALL BADGING.

MIRROR COVER FINISHED IN TEXTURED MATERIAL TO MATCH OTHER DETAILS

SATIN ANODIZED FINISH.

BALL-N-SOCKET CONNECTION

SEC.

"JEEP" GRAPHIC RECESSED 3 MM INTO SURFACE OF MIRROR COVER.

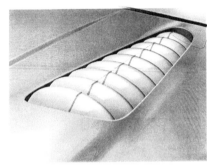

SEC.

VENTS TO BE FINISHED IN ANODIZED SATIN ALUMINUM.

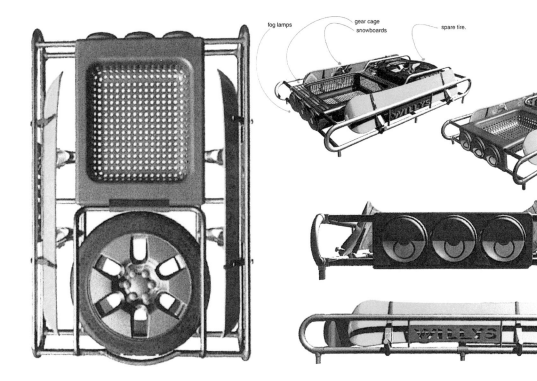

fog lamps

gear cage
snowboards

spare tire.

◁◁
**8.17** Jeep Willys concept, excerpts from the build sketch book by Jordan Meadows, featuring rearview mirror and hood vent details
SOURCE: IMAGE CREDIT, FCA/JEEP DESIGN.

◁
**8.18** Jeep Willys concept, exterior rear view, seen here in open top configuration featuring lightweight aluminum frame construction made visible through the rollbars and rear compartment tie-down rails
SOURCE: IMAGE CREDIT, FCA/JEEP DESIGN.

△△
**8.19** Jeep Willys concept, excerpts from the build sketch book by Jordan Meadows, featuring multifunctional roof rack system
SOURCE: IMAGE CREDIT, FCA/JEEP DESIGN.

△
**8.20** Jeep Willys concept, exterior detail seen here featuring gray tires. Building on the core value of treading lightly, the gray tire concept allows the user to experience nature without leaving unsightly dark rubber marks on it.
SOURCE: IMAGE CREDIT, FCA/JEEP DESIGN.

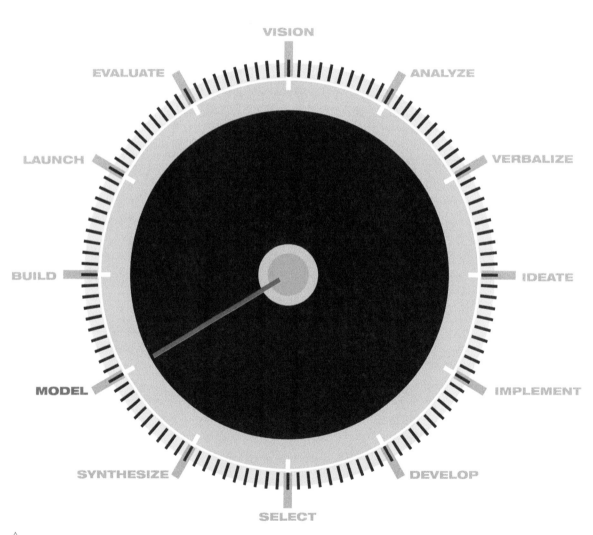

△
**9.0** Process locator gauge
SOURCE: IMAGE CREDIT, JORDAN MEADOWS.

**CHAPTER 9**

# MODEL
## Making It Happen and Translation to Digital and 3-D

**Making It Happen!**
One should never underestimate the power and impression a beautiful object gives when experienced in person. Ultimately our goal as designers is to create a meaningful experience for a user. The centerpiece of that experience is the object. It is the means for delivering the feelings and ideas of its creator to its user. To put it bluntly, it's a physical **3-D** world we live in, and designers are the arbiters of taste and judgment with regard to the objects we experience in it.

In the previous chapters of this book we have spoken about the synthesis and refinement of the chosen theme. Having then defined a direction, we also discussed how multiple work streams with regard to the interior, exterior, and the **UX** can be integrated. In this chapter we will backtrack slightly to more deeply understand the idea of how specifically concepts become a tangible 3-D reality. Various forms of model-making appear throughout the design process; we are concentrating on them at this point, however, as it generally proves most relevant once intention has been defined (see Figure 9.0.)

The design process is a series of decisions made in a sequential way to achieve a desired goal and objective. Each step along the way has meaning and importance. Three-dimensional models are essential to the creative process, aiding in the decision-making necessary to advance a design. Model-making is critical to validating an idea and gaining an understanding of how a theoretical concept can be applied in a real-world setting. This step of the journey is incredibly important as it is the key point where all the thoughts, ideas, imagery, and sketch-work generated are transferred into a three-dimensional reality.

Trained and talented professionals can dedicate an entire career to the craft. This chapter covers a broad overview of the different types most relevant to transportation and vehicle design. And while model-making is not the primary focus of most designers, it's crucial to understand the basics of how an image becomes a three-dimensional reality. This chapter will also cover deployment strategies for different types of model-making, providing insight into when and how they should be used.

J. Meadows

△
**9.1** MetalBack Café Racer, digital sketch
model, seen here blending wireframe
data and semi-rendered surfaces
SOURCE: IMAGE CREDIT, JORDAN MEADOWS.

▽
**9.2** MetalBack Café Racer, digital sketch
model, seen here in orthographic views
SOURCE: IMAGE CREDIT, JORDAN MEADOWS.

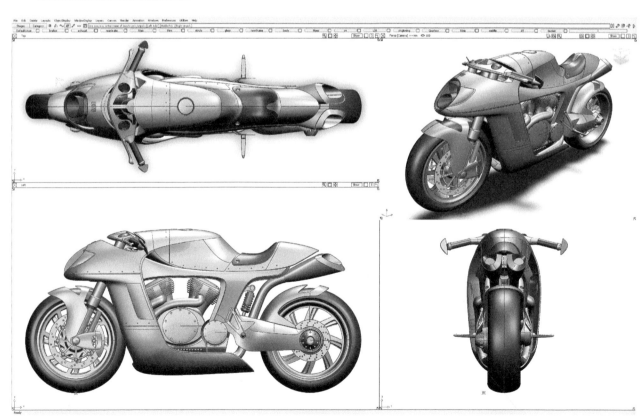

**MODEL**
MAKING IT HAPPEN AND TRANSLATION TO DIGITAL AND 3-D

### Virtual 3-D and the Digital Design Process

Around the globe, **transportation design** studios rely on **digital design** tools as a key part of the model development process. This was not always the case. Through the development of the discipline in most of the past century, vehicle design studios relied primarily on hard modeling and clay "actual models." Needless to say, the production of **hard models** was very time-consuming and often quite arduous. With the advent of digital technology, much of the development process is now handled virtually on-screen (see Figures 9.1 and 9.2). This has reduced much of the development time and cost in bringing a vehicle to market. These very sophisticated programs create

mathematical data to define the height, width, and depth coordinates of a virtual three-dimensional surface. Based on the Cartesian coordinates system, all modeling, be it virtual or analog, can be defined in XYZ points relative to a given origin. Understanding the Cartesian coordinates system is crucial for any designer in creating and communicating their project's intended form and geometry (see Figures 9.3 and 9.4).

There are various digital design programs, most of which can be grouped into three main areas of focus. The first, very user-friendly and well suited to broad-based rapid sketch modeling, programs such as Modo deliver relatively unrefined data for

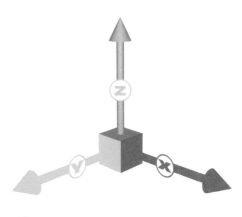

△
**9.3** The XYZ planes of the Cartesian coordinates system, an essential tool for all digital and analog model-making
SOURCE: IMAGE CREDIT, JORDAN MEADOWS.

▽
**9.4** MetalBack Café Racer, digital sketch-model, seen here with digital tape drawing used to construct the basic surfaces
SOURCE: IMAGE CREDIT, JORDAN MEADOWS.

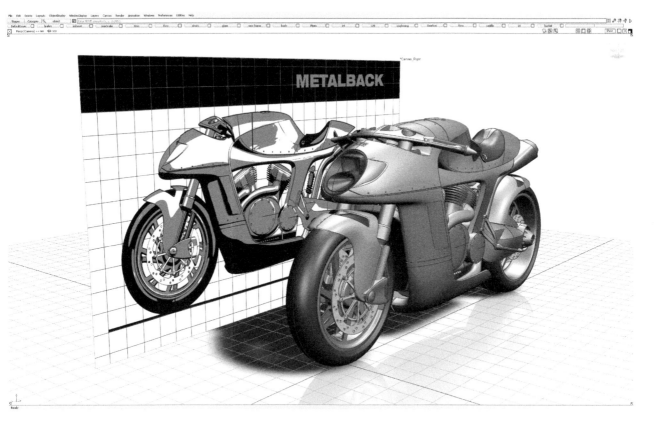

basic analyses. These can be used early on to establish basic volumes that reflect rudimentary forms and approximate design intent. These programs are also primarily used by designers in the studio environment and are quite effective in quickly roughing out an idea. There is a midrange of programs such as Alias and Maya that are used by both designers and digital **modelers**. This range of programs is not only well suited to sketch modeling but also offers a good degree of surface refinement capability. These midrange programs also have powerful rendering plug-ins that are well suited to design-based presentations. Having a good understanding of these midrange 3-D development programs can supercharge the designer's efforts. The third tier of digital modeling software is dedicated to engineering feasibility, solids modeling, and final surface release. These very dedicated programs such as Catia and Isem Surf are used primarily by **engineers**. They offer a high degree of tolerance and

control; however, they often lack the visualization capabilities of the midrange programs. When used effectively, all three are very well suited to provide excellent digital models in the place of costly 3-D models and prototypes (see Figures 9.5–9.8). The rendered MetalBack Café Racer in these figures was modeled using Alias Software.

△

**9.5** MetalBack Café Racer, digital model, rear three-quarter view, rendered with oxidized surfaces indicating recycled metal
SOURCE: IMAGE CREDIT, JORDAN MEADOWS.

▽

**9.6** MetalBack Café Racer, digital model, side and front view composition, rendered with oxidized surfaces indicating recycled metal
SOURCE: IMAGE CREDIT, JORDAN MEADOWS.

△△
**9.7** MetalBack Café Racer, digital model, front three-quarter view, rendered with oxidized surfaces indicating recycled metal
SOURCE: IMAGE CREDIT, JORDAN MEADOWS.

△
**9.8** MetalBack Café Racer, digital model, upper side view, rendered with oxidized surfaces indicating recycled metal
SOURCE: IMAGE CREDIT, JORDAN MEADOWS.

Alias Studio Tools is a common software package used by designers and modelers in the automotive and transportation design industry. It generates surfaces using **non-uniform rational basis splines (NURBS)** surface construction, based on mathematical splines and curves, and is in fact an outgrowth of traditional analog vehicle modeling techniques. When designers, modelers, and **fabricators** could not use basic geometric methods such as rulers and compasses to create a shape, they employed semi-flexible strips of wood or extruded aluminum splines. The splines were held at predetermined points by pins known as ducks. The ducks could be adjusted to affect the curvature of the spline, resulting in alterations of shape. Software developers sought to emulate this means of controlling the curve when developing the theory behind NURBS. Rather than having analog ducks, a designer or modeler can adjust the quality of the curve using digital control points. With this method, NURBS modeling allows for ease of control over the character of a particular curve or section. From a conceptual standpoint, the idea of using a curve to create a surface also connects well with the drawing process that most designers are familiar with. In fact, there are great similarities between working to a given tape line on a tape drawing or **clay model**, and constructing the key curves of the model and composing digital splines. The tool also works well on small objects such as interior components, cars and exteriors, and of course motorcycles (see Figures 9.9–9.11).

◁◁
**9.9** MetalBack Café Racer, engine detail, rendered with oxidized surfaces indicating recycled metal
SOURCE: IMAGE CREDIT, JORDAN MEADOWS.

◁
**9.10** MetalBack Café Racer, side view image, rendered with oxidized surfaces indicating recycled metal
SOURCE: IMAGE CREDIT, JORDAN MEADOWS.

▷▷
**9.11** MetalBack Café Racer, side quarter image, rendered with oxidized surfaces indicating recycled metal
SOURCE: IMAGE CREDIT, JORDAN MEADOWS.

▷
**9.12** MetalBack Café Racer, tank and steering column detail, rendered with oxidized surfaces indicating recycled metal
SOURCE: IMAGE CREDIT, JORDAN MEADOWS.

J. Meadows

J. Meadows

An alternate method for constructing surfaces digitally is based on polygon modeling. Other software packages such as Maya provide powerful capability for modeling surfaces this way. Rather than generating a network of curves and splines, this method of digital modeling is developed from mesh data created with polygons. Polygon models are a collection of vertices, edges, and faces organized to approximate and represent a surface. Whereas NURBS-based principles were developed as an outgrowth from old-world shipbuilding and coachbuilding with splines, polygon models are more akin to figurative sculpting. The principles of such are derived from model-making techniques developed in the entertainment industry to create surfaces with immense nuance and detail. However, they do have a difficult time representing a curved surface with typical automotive refinement.

Because polygon models use a different type of math, they are well suited to applications that are sensitive to data size. For example, most digital models used in video games and feature-length special-effects sequences are constructed using polygons. In the automotive industry, when scanning a soft-trim surface that has been sculpted with clay by hand, quite often polygon mesh is used to capture the data. Polygon models tend to be well suited for very fluid, organic, and soft feeling shapes. When executed by a skilled digital sculptor, this method of surface creation can be an effective way to generate informative sketch models early on in the design process. Both methods respond well to the mapping of surface textures and details as illustrated

here with the MetalBack Café Racer (see Figure 9.12).

### Digital Sketch Modeling

Many designers can use programs such as Alias quite early on in the process to rough out ideas and verify that they work over a given package. This process is referred to as sketch modeling, meaning it blends the **rapid visualization** sketching can provide with three-dimensional verification that modeling ensures. It can be a valuable way to communicate that a given theme will actually work in three dimensions under various conditions. This process begins with an initial side view drawn with fairly little perspective. Ideally a side view should have the clarity and information to the level of a traditional tape drawing over a package. The designer can then obtain a data file containing the hard points of the package. This would include occupant location, an outline of the **powertrain**, the wheels and tires, and any other required and given guide points such as bumper beams, glass planes, windshield, touchdown, etc. The designer can then arrange the isometric views that they have produced that defined the front, side, plan view, and rear. These planes will then be the basis for the key curves and splines laid out in XYZ coordinates.

Empathic quartering view drawings are quite compelling in selling a theme. They do a great job of getting everyone excited and interested in how the stance of the vehicle is communicated in the mood it portrays. However, fairly realistic isometric drawings are essential to creating a model and transferring the idea into reality. Most automobiles are read in side view predominately. Motorcycles

are similar in that much of the design can be communicated in side view. For this reason, having a compelling side view of your idea can lead and inform the sketch model process. Once the side view is finished and you are completely in love with the way it looks, you're ready to begin!

The basic rule of thumb to all modeling and sculpting is always begin with the largest elements first. It's helpful to have a plan of attack that addresses the first read dominant items initially, subordinate elements after, and finishes up with supporting details. Starting from your side view, capture the **silhouette** with splines. Using the various curve tools provided, trace over your side view drawing. Then capture the daylight openings, wheel openings, and main graphic features. Finally, use splines and curves to indicate subordinate details and transitions in body shape and section.

Then repeat this process in the front view, plan view, and rear view. Once blended together, one can then see a three-dimensional impression of how these key features and curves are relating to each other. This can then be used as an underlay for Photoshop work or two-dimensional rendering.

Continue adapting the key curves and sections to work with each other in all views. Time permitting, one can then construct some basic surfaces to get an impression of how the transition of forms may play out. Almost all vehicles are governed by the main structural elements contained within them. Earlier on in this book we established that the occupant package was part of this equation.

With automobiles the glass planes indicate the cabin and where the occupant package is most crucial. Most automobiles have a barrel-shaped glass that retracts into the door. The plan view curvature is governed by the beltline of the vehicle. The end view curvature is defined by the drop of the glass along with the "tumble home," which commonly refers to the angle that the side glass sits at in end view. This feature can be very important to establishing the visual center of gravity in the car. Cars with rails or very wide roofs in end view will read as very boxy. For example, dedicated off-road vehicles like the Land Rover Defender, or Jeep Wrangler, or Mercedes G Wagon tend to have very little tumble home and very flat end view sections. This is done to move the occupants as far out in end view as possible so they have the best command of road when doing extreme off-road driving, providing superior vision angles.

Conversely, sports cars that need to have a reduced frontal area in end view for **aerodynamic** reasons tend to have a good amount of tumble home. In either case, establishing the glass planes in side view will govern a good deal of the look and feel of your sketch model. Similarly, the angle of the windshield will also have a strong effect on the impression of speed and the feeling of your vehicle. When modeling, even in sketch form, this needs to be established fairly soon as it's a major compositional element. Two key features govern the windshield angle, the first is the windshield touchdown or **cowl point**. This is a result of many factors, primarily being where the engine box and firewall are located. This is a key

structural element of the front end of the vehicle. The engine and transmission are mated to the front of it, and the interior trim panel is mated to the rear of it.

The second crucial dimension when modeling a vehicle is to understand where the "header" is. This simply refers to the area of the vehicle where the windshield stops and the roof begins. From a packaging standpoint this can be a crucial feature as it defines many of the vision angles and entry and egress points. The third area for the glass planes that enclose the cabin is the backlight or rear window. Typically, there is much more freedom with this element. In any case, when sketch modeling, it helps to rough in the glass planes with a barrel shape on the side, and the desired windshield angle in side view. Another key aspect of the windshield is the plan view of the glass. This can establish where the A-pillars sit. Though almost all vehicles are composed in direct side view and the designer must predetermine the Y zero or centerline or silhouette section, the viewer tends to read the outside periphery of the shapes so the rails and pillars and elements further outboard tend to have more visual significance. This is the reason why a daylight opening or side glass shape can be such a key feature in the **aesthetic** read of the vehicle.

Once the glass planes have been roughed in, one can then continue the same process of laying out key surfaces between the main curves and splines that you've composed. Just having a basic hood surface, body side section, and some indication of the rear end can indicate a lot of how the vehicle will take shape in 3-D. Once having these in place, one

can then transfer back into 2-D and sketch over the rough model, indicating how the surfaces could potentially be blended together and how the transitions of the corners might be achieved. Having completed this and informed the next steps in a 2-D perspective drawing, one can then transfer back into 3-D and continue on with the same process of roughing in volumes and surfaces for the corners and transitions among the dominant elements. At this point, finessing the details can take some time. One has to judge whether it would be faster to complete a model in data or continue with the two-dimensional rendering of the design intent. In either case, the sketch model has served its purpose in that the theme has made the initial steps of transferring from a two-dimensional proposal to a three-dimensional object.

Alternatively, one can generate a sketch model using polygon modeling tools. Maya is a popular program that allows for this method. The end goal and steps are relatively the same as indicated above; to arrive at a rudimentary three-dimensional object that verifies that the proposal can be achieved in three dimensions. With polygon modeling, however, one would not start with key curves. And rather than splines, one would use control vertices to alter the shape. The initial first step is exactly the same. First, one needs to assemble isometric views to define what the object looks like in X, Y, and Z. One then begins to generate a single polygonal surface, adding subdivisions to create the height, width, and length of the object. With this process one continues to add surface subdivisions to capture the main compositional elements of the volume. Similar to a NURBS technique, the glass planes, roof, and windshield need to be defined. Having a mathematically correct barrel glass using polygon modeling can be a challenge. For this reason, one can approximate, based on an ideal section, plan view, and end view. Continue by adding polygons to capture the main side view features and wheel openings. Then repeat the steps of adding additional polygons to the front and rear of the vehicle to indicate the main design elements that need to be communicated. The aim of this, however, is still not to produce a final finished model, but rather to give oneself a good base in 3-D to sketch over, all to validate that the theme can be achieved in three dimensions.

Once an ideal perspective of your digital sketch model has been selected, transfer it back into 2-D. Using Photoshop or any two-dimensional painting or drawing program, one can then add the subordinate details that would be too time-consuming to build at this point. Examples would be wheels, tires, grill textures, lighting, and bright work features. Time permitting, one can then transfer back into 3-D and add a further round of polygons to the mesh to capture the subordinate details and design features.

### Three-Dimensional Data Development

Using an iterative process of two-dimensional and three-dimensional sketching can be an effective way of capturing the basic theme in data. One can then make a strategic decision, based on timing and resources, about how best to proceed. If the goal is to have an actual three-dimensional object that you can touch, tape on, and evaluate, then one should consider cutting a **scale model** with a **CNC (computer numerical controlled)** milling machine. The basic data files will need to be transferred out of Alias or Maya. Using a different program, the tool paths for the milling machine will have to be coordinated, and the choice will have to be made to cut a clay model or a density foam model. Clay gives the opportunity for further updates and to continue the design and refinement process. Foam is cheaper, and allows one to rapidly evaluate the work that's been done.

In any case, the next round of data development is to add a level of detail and refinement to the design. Nuance and organization of the shapes and transitions become ever more important. For this reason, NURBS-based modeling with splines tends to be a more popular choice as it affords a level of refinement in the execution of the surfaces. For enhanced three-dimensional data development, the basic principles can be continued from sketch modeling. However, experienced digital modelers will spend much more time developing a strategy for executing the model.

In addition to this, much more time will be spent evaluating both the main surfaces and transitions in areas of the car that are typically complex. In almost all cases these involve the corners of the vehicle. This stage of the modeling process will also involve a good amount of time evaluating the surfaces with different finishes and under different lighting conditions. With the exteriors of automobiles, there is generally a painted, glossy finish to be considered. This means that controlling the highlights and reflections will be an issue. To focus on this, most digital modeling programs offer several different evaluation tools so the operator can simulate and understand exactly how various reflections would play out over a given surface. Finally, most design teams will spend a good amount of time during the three-dimensional data development process viewing the virtual model at one-to-one scale on a projector or large screen monitor. This allows the designer to more accurately judge the character nuance and detail of a particular shape than can be done on a small desktop screen.

The three-dimensional data development process for interior design on vehicles can be even more intense. Most contemporary vehicles have a good amount of carryover components such as gauge clusters, ventilation registers, and **HMI (human–machine interface)** controls. All of these need to be integrated into the digital model being constructed for the theme. Because of the amount of work required, digital modelers can work in teams. And due to the nature of the medium being a non-destructive editing process, data development teams can generate several different proposals and evaluate one versus the next quite easily. Once a selection has been made, the data can then be exported for engineering evaluation. This would involve a process of transferring an Alias wireframe file, for example, into a program such as Catia. Engineers can then begin a lengthy process of validating and investigating the feasibility of a given design proposal.

## Rapid Validation Mock-ups

One of the main goals of the modeling process is to understand and communicate the three-dimensional spatial relationships of shapes and volumes. This goes for getting a feel for some basics such as, how wide does the vehicle need to be to feel stable? Or how large should a grill shape be before overpowering the other front graphics? For elements on a vehicle that can be represented in a two-dimensional way, designers often use cartoons or mock-ups to indicate how a feature or an element could be viewed on the vehicle. For example, exterior designers will typically print side view images of different wheel designs in different sizes and place them on the tires of the scale model or one-to-one model to get a sense of how they read. If the front end of the vehicle is flat enough, then one can print different grill shapes and sizes and quite literally pin them onto the model to understand the implications to the overall theme. These two dimensional cutouts are known as cartoons. And when photographed on the model, quite often they can be very convincing.

Using rapid validation mock-ups can help one to understand the interplay of shapes and volumes to a particular aesthetic; however, this method can also be very effective in understanding and communicating how the vehicle's riders and passengers interact with the proposal. Proxemics is the study of spatial relationships with regard to human social interaction. Anthropologists and psychologists use it to understand the impact of nonverbal communication. Based on the idea of a range of radiating spaces, from the intimate, to the personal, to the social, to the public, one can then break down and codify different behavior and interaction among people. Proxemics is also valuable for designers to understand. Many vehicles are social spaces in which the occupants can be closely placed together. To achieve this, designers use a combination of seating bucks and foam core mock-ups of the vehicle and packaging attributes in one-to-one scale. These mock-ups are essential to gain an understanding of how the occupants relate to each other within the vehicle and also how they relate to the vehicle itself.

Most design development teams will re-create the vehicle using what is known as a **whalebone buck**. Designers, fabricators, and modelers will typically use the digital data to cut sections of the vehicle in given increments. These can be 100–200 mm or any given dimension to communicate the basics of the shape. The sections are then printed out in one-to-one scale and used to cut foam core templates. The templates are then assembled in two of the X, Y, and Z axes typically. This effectively creates a skeleton-like impression of the actual vehicle. Evaluators can then sit inside this skeleton-like cage to get a feel for the key attributes of the package and the design proposal. In a family car or minivan, special relationships and proxemics can be crucial to its success. With a bit of imagination every aspect is available for review and judgment. Is the beltline too low or too high? Does one feel empowered while driving, does the vehicle have sightlines that encourage a command-of-road feeling? Does the occupant package feel crowded or spacious? Is the vehicle easy or difficult to enter and exit? Nearly every aspect of the user interaction except for the actual driving experience can be simulated with rapid validation mock-ups in one-to-one form presented in the **seating buck** or whalebone mock-up.

If time permits, a design team can go even further and combine milled foam portions of the interior design. Teams often do the same for the exterior (see Figures 9.13 and 9.14). These, combined with a steering wheel and seats, can communicate much of what it will be like to sit inside the finished vehicle. Conversely, on the exterior, rapid validation mock-ups can also be very helpful in understanding how various components relate to the overall vehicle. For example, on a one-to-one clay model, it can be helpful to produce a quick mock-up of a mirror or a door handle. These items can be placed on the model. One can then validate, is the door handle in the proper location? Is it too high for children to reach? Is the mirror too large? And is the aesthetic of these items relating well to the overall vehicle?

It is important to note that the goal of these models is to quickly mock up the idea and validate that the proposal will achieve its intended goal. It is important to focus on the basic function and evaluate its merits while the aesthetic is being refined. Rapid validation mock-ups are intended to achieve just that: quick visualization and verification that the proposal that you are pursuing will work as an actual three-dimensional object. They are effectively low-fidelity props to preview what's to come with the final composition.

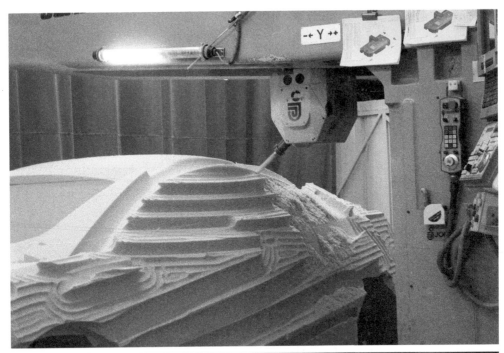

△△
**9.13** Rapid validation foam model, being milled in one-to-one scale. Note indicator on milling machine defining (Y) plane.
SOURCE: IMAGE CREDIT, FORD DESIGN.

△
**9.14** Ford Start concept, rapid validation mock-up illustrating seating configuration and occupant package
SOURCE: IMAGE CREDIT, FORD DESIGN.

# Clay Modeling
## Mazda Kiora Concept

△
**9.15** Mazda Kiora concept, scale clay
model, front three-quarter view
SOURCE: IMAGE CREDIT, MAZDA DESIGN.

Of all the modeling techniques
and methodologies used in a
contemporary vehicle design
studio, clay modeling is perhaps
the most fascinating. Becoming
a master sculptor can take years.
Using one's hands to guide very
rudimentary tools to create precise
and accurate surfaces in clay is
truly a remarkable skill. Doing
it at one-to-one scale on a very
large vehicle, repeating results
on a day-to-day basis within a

tolerance of millimeters, is even
more of a challenge. And on top
of all this, the ability to interpret
a designer's sketch, execute it in
three dimensions, and capture
the emotional impact, all while
honoring the engineering package,
is truly a craft bordering on
sorcery (see Figure 9.15).

Originally, automotive modelers
and fabricators created metal
and plywood armatures. These

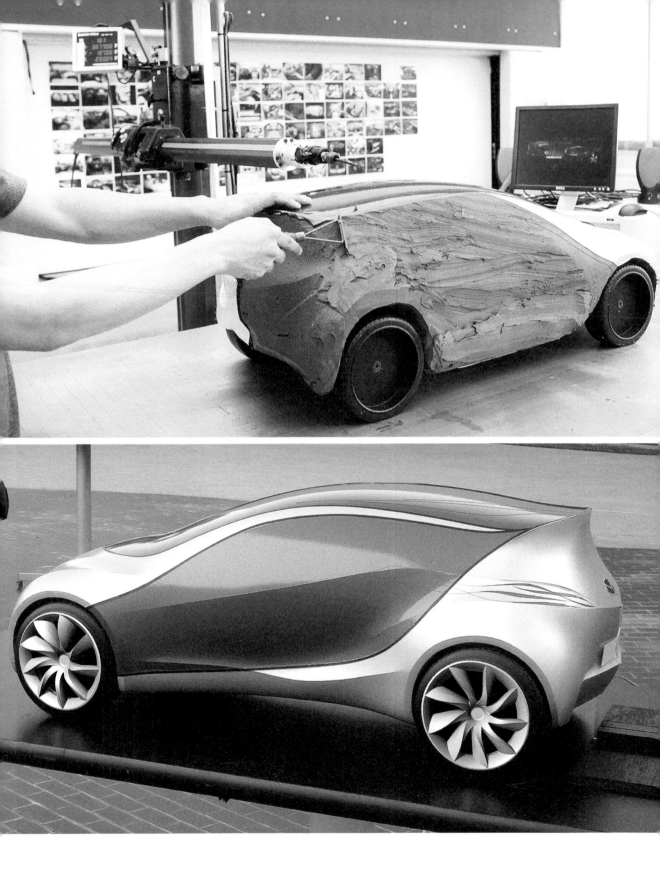

**MODEL**
MAKING IT HAPPEN AND TRANSLATION TO DIGITAL AND 3-D

armatures were then covered with plaster. The basic shape was governed by boat-like sections cut from templates of wood predetermined by splines. The craftsman would then carve away the plaster to reveal the final shape. Additional plaster could be mixed, added, and re-carved to alter the form. This yielded effectively a hard model. From that, measurements could be made and tools could be created. Needless to say, the entire process was extremely labor-intensive, time-consuming, and oriented toward producing one car at a time very slowly. During this period automobiles were low-volume artistic statements of opulence for the wealthy. Harley Earl was credited with changing this process. While working as an independent in Southern California in the mid-1920s, his studio pioneered the shift away from plaster to clay as a much faster way of achieving shapes in 3-D. Earl was recruited to become the head of General Motors' art and color department. This

was effectively the birthplace of contemporary automotive design. Being faced with multiple projects running concurrently, the method of clay modeling was perfect for the requirements of the quickly growing organization looking to mobilize a nation.

Many of the early principles of clay modeling exist to this day. Sculptors begin with a metal and wood armature typically. The wheelbase and track are adjustable on these armatures to accommodate changes as needed. They are then packed with density foam that is cut over and milled to within a predetermined dimension of the final surface. The model is then packed with special automotive clay formulated for dimensional fidelity. It's typically denser than what you'll find in your local art shop. From this point the decision is made to pursue one of two processes. With a more traditional process, a clay modeler will produce basic templates from a tape drawing indicating key sections. These

include a centerline, a cross-section through the wheelhouse openings, a cross-section typically through the occupant **H-points**, and a plan view section through the wheel centers. The modeler will then pack the clay to the basic templates, and then when the basic proportion is ready, use sweeps tools and splines to capture the impression from the designer sketches. With a more contemporary process, these rough packs of clay are then cut with the CNC machine using two types of tool tips. The cuts usually fall between three axes and five axes and are performed with a flat head or ball-tipped tool. These tools leave a small texture. Clay modelers can then take the mill surface and begin to work their magic (see Figures 9.16–9.18).

In almost all cases there is an initial impact of seeing the idea rendered in three dimensions. Quite often, the impression is different from the two-dimensional work. The first step the designers and modelers will take is to get a

◁◁◁◁◁
**9.19** Mazda Kiora concept, full-sized clay model being evaluated in outdoor review yard with BMW Mini as segment scale reference
SOURCE: IMAGE CREDIT, MAZDA DESIGN.

◁◁◁◁
**9.20** Mazda Kiora concept, full-sized clay model being evaluated in outdoor review yard. Note the wheel design is actually a two-dimensional print done as a rapid validation mock-up.
SOURCE: IMAGE CREDIT, MAZDA DESIGN.

◁◁◁
**9.21** Mazda Kiora concept, full-sized clay model being developed with tape lines indicating changes to the key features
SOURCE: IMAGE CREDIT, MAZDA DESIGN.

◁◁
**9.22** Mazda Kiora concept, full-sized clay model being developed with tape lines to indicate changes to the key features, seen here with BMW Mini as reference for scale. Outdoor tape sessions and segment comparisons are essential to the development of any full-sized model.
SOURCE: IMAGE CREDIT, MAZDA DESIGN.

◁
**9.23** Mazda Kiora concept, full-sized clay model being evaluated in outdoor review yard fully dynoced. Outside reviews are essential for getting a long-distance view of the design.
SOURCE: IMAGE CREDIT, MAZDA DESIGN.

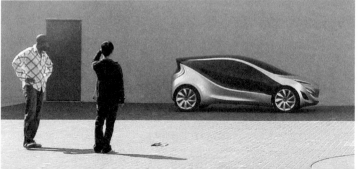

long view of the model in bright conditions. This is normally done in an outdoor review yard where evaluators can view the design from a distance as well as up close (see Figures 9.19–9.23).

The next step is to then establish a list of large changes that need to be executed. These would involve alterations to the main mass of the vehicle if needed, the roofline with track, etc. The truth is that vehicles have a physicality to them. Shape and volume of this size and mass can only truly be appreciated in one-to-one scale. And it's not until this point that a designer and modeler, or

△△
**9.24** Mazda Kiora concept, Jordan
Meadows, design manager, and Mickael
Loyer, principal designer, with full-sized
clay model being evaluated in outdoor
review yard fully dynoced
SOURCE: IMAGE CREDIT, MAZDA DESIGN.

△
**9.25** Mazda Kiora concept, full-sized
clay model with full dynoc for surface
evaluation
SOURCE: IMAGE CREDIT, MAZDA DESIGN.

anyone else who is evaluating the proposal, can truly get a feel for whether it's working or not. After this evaluation the clay modelers will then make the adjustments. The surfaces are refined and then typically another review will be conducted. This time the clay modelers will wrap the refined surface with a slick ultra-thin plastic material known as dynoc. This material is almost always some sort of silver or neutral for the sheet metal elements and a dark gray or black for the glass elements. When applied, it can give an indication of the graphic breakup of the vehicle and allow one to read the highlights and transitions of reflections. Designers will also add tape lines for the part breakups, door shuts, hood and bonnet cuts, etc. At this point, photographs of the model effectively look like a prototype manufactured vehicle. One can then use these images to do overlay renderings that communicate further updates and enhancements to the design. The clay modelers will then strip the dynoc, do another round of surface development, and repeat the process as many times as the schedule will allow to achieve the final iteration. This interaction of designer and clay modeler studying, interrogating, and evolving every last nuance and detail of the vehicle's surface is the core of how a design makes the transition from two-dimensional proposals to three-dimensional reality (see Figures 9.24–9.27).

△△
**9.26** Mazda Kiora concept, scale and full-sized with full dynoc for surface evaluation. Note the two full-sized digitizers to record the design in XYZ coordinates.
SOURCE: IMAGE CREDIT, MAZDA DESIGN.

△
**9.27** Mazda Kiora concept, full-sized clay model being evaluated in outdoor review yard fully dynoced. When complete, it can then proceed to a hard model prototype phase.
SOURCE: IMAGE CREDIT, MAZDA DESIGN.

## 3-D Printing, Rapid Prototyping, and Hard Model Fabrication

Most vehicle and transportation design studios employ hard model fabrication specialists along with clay modelers. This is a specific craft dedicated to the creation of objects that are not sculpted with clay, yet require a good degree of finesse and skill. Fabricators tend to be involved with every aspect of 3-D creation from the armature of models to the execution of every subcomponent of the vehicle. For example, roof racks, door handles, mirrors, and grill work can all be very important to completing the look of a proposal. One of the most important features beyond clay is the painting and finishing of the surface. When a clay model is complete, often the design team will make the decision to paint it, or in some cases cast it in fiberglass for durability in transportation and showings at different presentations. In either case it will be sealed with fillers, primers, etc., and a detailed process of sanding the surface will then take place. Finally, automotive grade paint and clear coat will be applied. A good model-maker will also spend time on the lamp areas. These can be quite detailed design statements in their own right. Typically, a hard modeler will take a plaster cast of the exterior surface executed by a clay modeler. Clear plastic will then be vacuum formed over the shape. The inner parabola can then be executed with clay or a hard material. Once all the constituent parts are complete and assembled, the proposal will look and feel very much like an actual vehicle. In fact, external evaluators who are not familiar with the model-making process can often mistake them for being ready for users to take a drive (see Figures 9.28–9.32).

Yet another method for verifying and communicating ideas in 3-D involves desktop prototyping. 3-D printers and rapid prototyping machines are a common means for transferring virtual files into actual objects. Most 3-D printing and rapid prototyping machines

△△
**9.28** Mazda Kiora Concept, side view of hard model prototype being developed
SOURCE: IMAGE CREDIT, MAZDA DESIGN.

△
**9.29** Mazda Kiora concept, detail view of hard model prototype being developed
SOURCE: IMAGE CREDIT, MAZDA DESIGN.

**9.30** Mazda Kiora concept, rear view of hard model prototype being developed
SOURCE: IMAGE CREDIT, MAZDA DESIGN.

**9.31** Mazda Kiora concept, rear fascia diffuser detail of hard model prototype
SOURCE: IMAGE CREDIT, MAZDA DESIGN.

**9.32** Mazda Kiora concept, final assembly of hard model prototype
SOURCE: IMAGE CREDIT, MAZDA DESIGN.

have limitations on scale and size. They are quite well suited for small objects and components. Interior parts such as shift knobs, switches, bezels, etc. are well suited for this process. Components for scale models of exterior proposals such as wheels, mirrors, and trim details are also easily created. One can quickly see that a designer who is quite fluent with Alias can generate sophisticated milled models and rapid prototypes, limited only by budget and schedule.

## Deployment Strategies for Different Types of Model-Making
The design process is very much a journey. Model-making is a crucial point of the process where stumbling blocks may rise and surprises occur. In any case, a design hero absolutely must engage in the challenging exercise of transferring vision, strategy, and 2-D imagery into the three-dimensional world! Many designers simply hand over their work to executioners. Quite frankly, this is very similar to making it halfway through the journey and deciding to go home. Generating ideas is only part of the challenge.

△
**9.33** Mazda Kiora concept, rear three-quarter view of vehicle
SOURCE: IMAGE CREDIT, MAZDA DESIGN.

Follow-through is where the adventure begins. And for those who are committed to the quest of actualizing an idea, it's important to understand the various techniques of transferring 2-D to 3-D. We have outlined some of the basic techniques in this chapter. Many professional modelers have their own methodologies and secrets to success. However, from a designer's standpoint, it's important to

know how to implement the type of model-making that offers the greatest verification during the development process. This can be based on many factors such as schedule, funding, and available resources. In almost any case, most strategies will involve iterative phases of work. This effectively forms a four-stage loop; the first stage of this loop begins with two-dimensional sketching and development, the second involves

virtual 3-D data development, the third involves milling or printing in 3-D in clay or foam, and the fourth involves handwork, adjustments, and human finesse. Most design studios will begin by executing the stages in reduced scale and then transfer to full size. They will then cycle through the stages as many times as the budget and schedule will allow. Each step of the way there should be a critique and evaluation by the

△
**9.34** Mazda Kiora concept, front three-quarter view of vehicle
SOURCE: IMAGE CREDIT, MAZDA DESIGN.

designers and key stakeholders. This iterative process of utilizing various modeling phases allows for verification and validation of an idea through many filters and different points of view. It gives the team a chance to evolve and enhance all the key aspects of how the user will interact with the object in real terms. Furthermore, it allows the designers to hone the aesthetic, ensuring that the final object appears exactly as intended (see Figures 9.33 and 9.34). In today's competitive landscape, each one of these four phases can be crucial to a product's success. Ask any veteran of the vehicle design game and consistently you will hear, understanding the modeling process is mission critical!

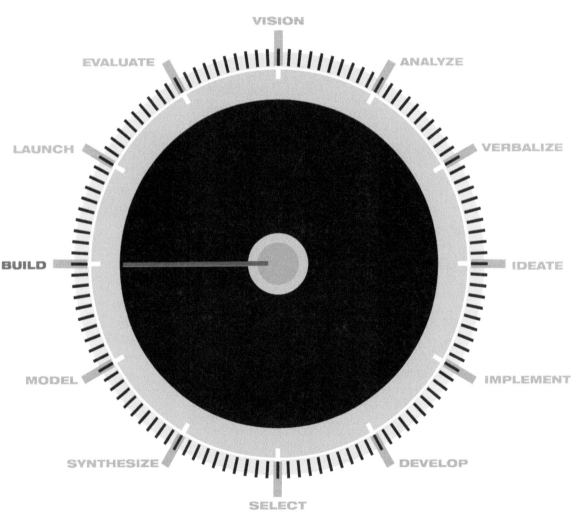

△
**10.0** Process locator gauge
SOURCE: IMAGE CREDIT, JORDAN MEADOWS.

**CHAPTER 10**

# BUILD

Becoming a Reality, Engineering, Mastering, and Sharing a Design Proposal with the Larger Product Development Community

## Vetting an Idea

At this point in the design journey the focus shifts to building a vehicle that moves both the heart and mind. The proposal is pretty well matured. You've generated several ideas and distilled them down to the strongest candidate. You made a wise choice and proved that it can make the trip from 2-D to 3-D. For many, this is the culmination of the creative process involving multiple work streams. In professional settings, however, we're not out of the woods yet; in fact, we're just about to head into the woods! (See Figure 10.0.)

Typically for production programs, there is a hand-off between the design studio and the broader product development community. Once the design direction has been confirmed and the various stakeholders are comfortable with how the project has evolved, a lengthy process of vetting the idea then begins. It goes without saying this stage of the process is hugely important. In fact, it is the crucial point that a vehicle develops a level of commercial credibility.

Good design moves the heart and the mind. It is the combination of **aesthetic principles** and pragmatic thought. But ultimately it must be usable. Beautiful ideas that never make it to the public have limited impact on the designer, and virtually none from a business perspective. Also, one cannot underestimate the importance of safety, durability, convenience and ease-of-use. In fact, this is what truly separates the professionals from the novice/ hopefuls. Usability in either function or appearance should never be presumed. It requires a good amount of diligent work to get a product to a point that the user can interact with it seamlessly and enjoy its intended function in a safe manner. And every designer who undertakes the task of creating vehicles should understand that a significant part of the journey and an enormous amount of work go into putting a vehicle into the public realm. Many creative studios have endless proposals that never made it. In professional settings, the kill rate can be very high. Almost every **transportation design** practice has archives full of lofty ideas and proposals that simply were not usable at the end of the day, or could not be made feasible. Passing through this gauntlet is no easy task. In fact, it's where some creative individuals actually meet the demon/dragon along the creative journey, and those who succeed are sure to emerge on the other side as true heroes of the process.

Most vetting of an idea done in concert with a broader product development community involves the creation of prototypes. This is the only way that an idea can truly be validated and made safe and usable. They are both learning tools and proof of project parameters. This chapter will cover a brief overview of the different types of prototypes, and their relevance to the respective parts of the process. In most cases these fall into three main categories: the first is typified by **hard models** and prototypes that prove out the visual aspect and aesthetic requirements of the vehicle. These serve to allow the designers and product development teams a chance to view how the object will appear under multiple conditions and in various applications. For example, a seating prototype that allows the designer to trim in multiple different materials and leathers would be valuable in seeing if the seats' appearance would potentially fit more than one vehicle. The second group of prototypes is done to prove out a particular function. These are commonly known as mules. In this case, the aim is not to evaluate the appearance, but rather to gauge whether the object is performing its intended function in the ideal way. The third group involves manufacturing feasibility prototypes. These typically combine appearance and functional attributes. However, the goal in their construction

is to prove out the optimal way that the vehicle can be mass-produced. Each one of these groups plays a very different role in the development process, yet all are essential to putting a vehicle on the road. And while a designer may not be personally responsible for them, it is essential that one has a grasp of exactly how they can impact their intended design.

### Engineering, Processing, and Testing

Vehicles need massive amounts of time and energy, and large groups of people, to mass-produce. Hence the financial and economic investment requires vetting and testing of an idea to make sure that the investment is a solid one. In short, the creation of a vehicle line for mass production can be a billion-dollar business wager that involves people's safety and well-being. For these reasons a good amount of processing and testing takes place to ensure a design is fit for the road.

The first group of concept testing and appearance evaluations can range from hard prototypes that allow the development teams to critically evaluate the **aesthetics** and basic functionality of a design. This can also involve digital simulators that can provide true-to-life and accurate re-creations of how a vehicle will be experienced. Many design studios use three-dimensional and virtual reality simulators when evaluating interior design in particular. In these cases, the usability of the **human–machine interface (HMI)**, particularly with regard to information and entertainment, is crucial. For example, if a team is evaluating a gauge cluster to understand what type of fonts are most legible when glancing

while driving, a simulator would be very helpful. Additionally, if a team is evaluating where a control should be placed so that it does not distract the driver, a simulator could be crucial.

Another important group of prototypes are dedicated specifically toward long-term durability, quality, and stress testing. Vehicles need to be able to perform in a wide range of outdoor applications. Manufacturers must ensure that their offerings will remain functional in extreme circumstances, such as subzero temperatures, foul weather, extreme heat, and many different types of road conditions. Added to this is the fact that the lifespan of most vehicles can last several years, if not decades. This can impact almost every aspect of the design. For example, if a new exciting color is proposed, the manufacturer would typically do long-term ultraviolet testing to ensure that the paint would remain not only vibrant and attractive, but also durable against corrosion and weather damage. To achieve a gorgeous new door shape or body side section, a particular type of hinge mechanism may be required to perform the necessary geometry to open the door effectively. If the mechanical engineering at this hinge is unique, it would need to be suitably tested for long-term durability. Consider the complexity of sliding glass, swinging stamped parts, organizing thousands of components, and sealing them all together with weatherproof tightness that lasts over decades! One can then imagine the countless hours of engineering coordination and developmental testing a vehicle must undergo.

Another group of functional prototypes is specifically created to evaluate safety (see Figure 10.1). Needless to say this is the most important aspect of the vehicle's design and engineering. In fact, every vehicle designer and creator has a moral obligation to provide the safest vehicle possible to the user. Obviously special exceptions such as motorcycles do exist; they are inherently unsafe. However, even in this case the basic design needs to adhere to general principles of what is relatively presumed to be a safe operation (see Figures 10.2 and 10.3).

Safety testing is crucial. Some brands such as Volvo have built an entire mission and culture to support this belief. In the mass production of most automobiles, designers need to provide ample accommodation for required safety devices such as impact beams, airbag provisions, safety belts and harnesses, and with exterior design, pedestrian protection. Added to this, many corporations have internal standards to align with their brand goals and objectives. Furthermore, these might correlate with insurance industry targets. All these factors require hours of testing to ensure that a design is performing safely while being used in a range of conditions.

Third, a comprehensive group of prototypes involves testing and vetting an idea for its manufacturing feasibility. Automotive manufacturing plants are hugely complex feats of engineering in and of themselves. In a sense, a plant is a fascinating machine designed specifically to create equally complex and fascinating machines. Added

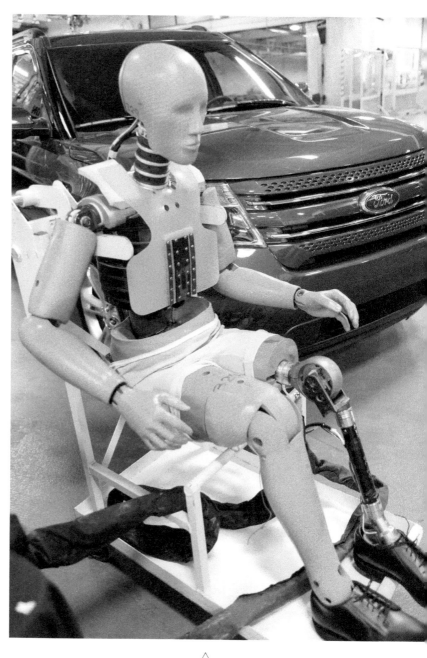

△
**10.1** Typical adult crash test dummy
SOURCE: IMAGE CREDIT, FORD.

to this, a feat of coordination is performed with remarkable speed and accuracy. Raw materials enter on one side, finished vehicles exit the other. In between, every square centimeter is choreographed and optimized to produce several cars

per minute amounting to hundreds of thousands per year, each comprising several thousand parts (see Figures 10.4 and 10.5).

Almost all manufacturers make extensive use of a pilot facility that tests the throughput of every component required to assemble a complete vehicle. Testing for feasibility requires a borderline obsessive attention to detail, as well as an incredible appreciation for the sequencing and coordination of the vehicle's manufacture. Getting any one point of this wrong can amount to not just one defect, but a defect that's repeated tens of thousands of times. Needless to say, it can quickly become very expensive for an organization. At this stage of the game a design can still be affected by what the plant can accommodate. Even the overall length of the vehicle can have implications on the assembly line, affecting speed, frequency, and overall output of the plant. Conversely in low-volume manufacturing scenarios, where a good degree of hand assembly is involved, manufacturing feasibility and ensuring flawless results are

◁◁
**10.2** Typical frontal impact crash test
SOURCE: IMAGE CREDIT, FORD.

◁
**10.3** Typical crash test dummies representing various human body types, weights, and sizes
SOURCE: IMAGE CREDIT, FORD.

▷▷
**10.4** Robotic spot welding mechanism in Tesla plant
SOURCE: IMAGE CREDIT, TESLA.

▷
**10.5** Robotic assembly mechanism
SOURCE: IMAGE CREDIT, TESLA.

equally as important. The luxury end of the spectrum requires just as much testing and vetting to ensure the highest quality product (see Figures 10.6 and 10.7).

## Market Research, Clinics, and Gauging Acceptance

As you can quickly see, due to the massive resources it takes to produce a vehicle, a good degree of research and testing is needed to ensure the investment will yield a profitable product that will be successful in the marketplace. And just as vehicle manufacturers spend a good amount of time vetting an idea from a technical standpoint, a lot of effort goes into understanding exactly what consumers will find desirable. Like many aspects of vehicle design and development, one can dedicate an entire career to understanding and mastering this stage of the process. When performed well by informed professionals, **market research** can yield some tremendously valuable insights.

However, there are dangers. If the research is poorly conducted, or misinterpreted at the wrong time, the information can be not only confusing but also very distracting and in some cases downright destructive.

Most market research is designed to yield qualitative and quantitative information about how consumers may potentially react to certain aspects of a proposed product or service. Having a combination of both scientifically compiled data and subjective impressions is the goal. Once a method of research has been decided, the effort is typically conducted in four steps. This begins with defining a sample or specific group of people to learn from. The second step is to gather data, collecting it typically from questionnaires for quantitative output, and discussions for qualitative findings. The third step is to process and analyze the data. Then the final step is for the development team to draw conclusions and formulate

assumptions based on findings. In a broad sense, the entire exercise is needed to see *why* and *how* consumers may react. The skill in interpreting the findings is to understand and learn from the *why* component. This illuminates the process and can lend context to product development decisions (see Figure 10.8).

Many organizations conduct market research around three specific stages in the development process. The first is exploratory and done early to understand the user, their wants, and needs. Much of this is covered in Chapter 2 where we discussed **archetypes** and user persona. This gives a basic overview and orientation for the program. The second phase of market research can sometimes be controversial. This involves asking respondents to rate or choose various design proposals. This typically is done after much of the initial theme development is completed and prior to hand-off to the broader engineering and

◁◁
**10.6** Robotic painting mechanism
SOURCE: IMAGE CREDIT, TESLA.

◁
**10.7** Tesla Model-S on assembly line
SOURCE: IMAGE CREDIT, TESLA.

△
**10.8** Market research attendants rating various images
SOURCE: IMAGE CREDIT, FORD DESIGN.

product development community. The third phase of market research is conducted very close to the end of the process in that respondents are asked to evaluate and confirm acceptance relative to its marketplace competitors. This can involve many objective components, such as price, features, and performance statistics. This phase can be valuable in assisting the product development team in finalizing how to present, advertise, and market the vehicle.

Again, there are both benefits and drawbacks that need to be considered. The fundamental danger is that respondents are asked to make a choice based on pre-existing assumptions usually derived from their current context. The reality is, however, no one can predict the future and the competitive marketplace may be very different at the time when the vehicle is launched. For example, if a design team is evaluating their proposal relative to cars that are on the market, this infers that the competitive dynamic will be the same moving forward. However, it is entirely possible that new competitors posing a different kind of threat may emerge.

Most market research also takes place in a controlled environment in a very short amount of time. It is simply not possible to allow respondents a chance to formulate an opinion over extended periods. Because of their complexity, many vehicles, however, require long-term evaluations for the user to truly appreciate their strong points. Added to this is the fact that most consumers are not visionaries. When forced to offer a rapid response, they tend to go with the familiar, in spite of the fact that

the new alternative may be better suited to their needs. In fact, there are many examples of highly successful products that did not do well in research. Conversely, there are as many examples of products that did very well in research, but went on to fail in the market.

On the plus side, market research can be a very valuable tool for establishing a broad general direction to orient a program and identify macro trends. It can also be very helpful in understanding what specific aspects of a vehicle may be polarizing or controversial. On the whole, when used to add context to decision-making, market research can be a valuable source of information.

## Early-Stage Vetting for Designers

It's quite easy for designers to presume that many of the topics covered in this chapter involve what happens primarily after most of the heavy creative lifting has occurred. One must generate a good idea and execute it before it can be vetted and tested? This is only partially true. The reality is there are some essential topics to be aware of as their implications can drastically affect the way that a final product is experienced. The design process (in the true meaning of the term) starts with the initial idea and runs all the way until the point the user engages with the vehicle. In keeping with this idea, there are some basic objectives to bear in mind relatively early on in the creative journey.

From a technical standpoint, it is very helpful to become familiar with pre-existing examples of successes and failures. A bit of early-stage technical

benchmarking is a valuable exercise. Once your design proposal has taken shape, whenever possible, use mockups and validation models in real-world scenarios. For example, say a design proposal calls for a significantly higher beltline or smaller rear window. Marking up and altering an existing interior and actually seeing how it feels while driving could be immensely informative. Another valuable vetting technique is to solicit varied opinions early in your creative process. Attempt to include an external input in critiques and reviews. Designers by nature have a creative point of view. It can be eye-opening to hear how your work may be viewed by someone who represents your intended user. To that end, attempt to share early and share often if a project is not secret. Internet and social networking sites can be a great way to collect multiple points of view and conduct informal research on how your project may be perceived. Always consider safety as one of your main priorities. Finally, remain rooted in plausible considerations and always cite examples when identifying the most progressive stretch targets. Understand that in the end it only helps if your project is predicated on proven rocket science rather than mysterious alien technology. Keeping these basics in mind early in the journey will help to ensure that your vehicle conquers all the various testing required to become a reality.

# Q&A

## RAJ NAIR

EXECUTIVE VICE PRESIDENT,
PRODUCT DEVELOPMENT, AND
CHIEF TECHNICAL OFFICER,
FORD MOTOR COMPANY

*Raj Nair is Executive
Vice President, Product
Development, and Chief
Technical Officer (CTO), Ford
Motor Company, effective
Dec. 1, 2015 (Figure 10.9).
In this role, Nair has global
responsibility for all aspects
of the company's design,
engineering, research, and
product development.*

*Nair joined Ford Motor
Company in 1987 as a Body
and Assembly Operations
launch engineer and has held
a number of senior positions
in Manufacturing, Product
Development, and Purchasing.
Besides North America,
Nair has also worked on
assignments in Europe, South
America, and Asia Pacific. Nair
holds a bachelor's degree in
Mechanical Engineering with
an automotive specialty from
Kettering University in Flint,
Michigan, and was named the
recipient of the 2012 Kettering
Alumni Award for Management
Achievement. In 2014, he was
named the Fortune Automotive
Businessperson of the Year. In
addition to his duties at Ford,
Nair serves on the board of
trustees of Kettering University.
Note: upon the time of
publishing Raj has assumed the
role of Executive Vice President
of Ford of North America.*

△
**10.9** Raj Nair, Senior VP of Product Development,
Ford Motor Company
SOURCE: IMAGE CREDIT, FORD.

**Question 1: Who are your
personal design and
engineering heroes? Who do
you most admire, and why? How
does this key influence relate to
the work you do for Ford Motor
Company?**

▶ Hero is a strong word, so I try
to recognize from my time with
the company that so many things
we do require a team effort at
the end of the day. It's very rare
that anything of any impact or
greatness in society or history is
done by an individual, even though

an individual may get the credit for
it. So there are certainly aspects
of certain cars that I grew up with
and admire that stuck with me.
Cars like the Dino, the Daytona,
Pininfarina's work and Fioravanti,
as a leader. The Lotus Esprit and
Guigaro's work, of course, the
Muira and Countach, although I
can't stand the look of that one
now, but generally Gandini's
work. And, as for overall design,
certainly Dieter Rams' principles
have had an effect on me. From an
engineering perspective, in terms

of pure engineering principles, certainly Colin Chapman, the boundaries he pushed. And for that matter, Gordon Murray and what he did. So, as a Formula One fan, I have to say Formula One engineers. Then once you get to production, it's very different and it's much more of a team effort. It may be the biggest team sport/ industry. So, certainly from an engineering perspective, and the management and running of a big organization, I learned a tremendous amount from Derrick Kuzak: my predecessor. Obviously I was very close to him. He mentored me, so I knew him the best. So in terms of how to run product development at Ford Motor Co. the guy I learned the most from was Derrick.

**Question 2: What type of strategies and techniques do you encourage when vetting an idea, validating new solutions, and testing a vehicle's potential for market success?**

▶  Relative to understanding a vehicle's potential for market success, it certainly depends on the vehicle. And so there are certainly aspects of established vehicles and segments that allow for some benchmarking, evaluating our own performance relative to competitors, etc. The depth of knowledge we have about our customers, particularly if it's a segment where we ourselves, to some extent, represent the target customer, allows us to use our own intuitive feel. And, for us at Ford, the two biggest examples of that are F-150 and Mustang. And although the market research is very useful, we know that product, and we know that customer so well that, to a large extent, we rely on our own thoughts and feelings about innovation, letting our

intuition guide what we feel about the aesthetic, or about the utility. I think for other segments where we ourselves don't necessarily represent the customer, it's about ensuring that we get the appropriate target market insight. And I don't necessarily think it's just about going out and doing a clinic. They are useful, but it's about understanding that the quantitative feedback sometimes isn't as valuable as the qualitative information you get. So that's a danger we have to caution ourselves on. The quantitative numbers are a great way to understand the relative positioning, but you need to get to the qualitative info to understand what drove that relative positioning. There's a myriad of tools available to us. The biggest danger I think for our business is to over-rely on one, and to think that whatever toolset used for the next F-150 is also going to work for the next Figo. Or for the next GT that is on the opposite end of the spectrum where there wasn't any market research. It was very much about designing for the purpose of certainly winning Le Mans, and basically for ourselves!

As for ideas and solutions, it depends what we're talking about. The nature of how white-space or how future-oriented it is, and who within the organization is going to be the subject matter expert on it. With the technical aspects, it can be difficult because there can be such varying views, even internal to the company, as to where people think things are headed. Where an approach has viability, versus where an approach will just lead to a dead end. And by the time you get the actual test results, it's too late. You had to make a decision prior

to that. So again some solutions lend themselves to very subjective aspects of the data, and others lend themselves to the faith in the teams and the subject matter experts. As for aesthetics, again, it's complex. There are aspects of this area I view similar to the music industry. If you're to look at the music business as an actual industry, you certainly have the creative process and could say you have the songwriters and those are designers. And often in studio settings you have professional musicians, who are like our modelers, be they clay or digital. And then you have the producers, who you could almost call our chief designers and managers. You also have the executive producers who to some extent don't have the talent to create, but they do have an ear for what may work or not. It's very similar to having an eye for what will work and what won't. But the other thing the executive producer has to do is have a good understanding of the market place to know whether it's going to be viable. And then they have to go convince the label that it's going to be viable and marketable. And in the automotive industry the label tends to be the rest of the corporation with the marketing and sales team in particular. Sometimes I view myself in that chair of the executive producer. I'm certainly not the one with the creative idea, or carving the clay or changing the curve. Nor am I involved in the day-to-day such as Moray or Freeman. But I'm the one having a reasonable eye, having an understanding of the market, and then going to convince the rest of the corporation that it's the right approach. That's the role I serve.

**10.10** Ford GT, competition car testing on banked racetrack
SOURCE: IMAGE CREDIT, FORD.

**Question 3: What role do motorsports and racing play in vetting an idea in the vehicle development process? How is it applicable to consumers who aren't enthusiasts at the end of the day?**

▶ Those are two separate questions and, to be honest, motorsports has gotten so specialized and mainstream and automotive engineering has gotten so specialized that the overlap is actually quite limited relative to actually vetting a technical solution. Many of the engineering tools we use, however, are very similar. So to vet a new engineering tool to prove

out a design is very applicable. For example, computational fluid dynamics, or stress analysis, kinematics, or combustion analysis, etc. Sometimes this is where it's interesting because they go back and forth, and the production group is actually ahead. Then we are able to apply that leverage in race engineering. But there are some tools where race engineering is ahead because they can prove out much faster, having results every weekend, whereas production takes a bit longer. So there is a lot of back and forth on engineering tools. The amount of actual technology you would find on

your car is limited. EcoBoost is probably the exception to that. The Ford GT as a vehicle is probably an exception to that. In fact, Ford GT is unique even amongst the vehicles it's racing against. It actually has things in racing that are also helping us in the road car just by virtue of the fact that we're developing it for production at the same time. That usually doesn't happen.

It's a great tag line, being born on the racetrack, but it's rarely true. But for this car it is, because the development is actually happening at the same time (see Figures 10.10 and 10.11).

△
**10.11** Ford GT, two competition cars high speed testing at racetrack
SOURCE: IMAGE CREDIT, FORD.

And then relative to the customer, they're certainly benefiting from those engineering tools and technologies, even though they may not be aware of what they're derived from, or where they were proven out. And certainly some of the high-level concepts of the EcoBoost engines and some of the actual 2017 model year parts that first saw the light of day on the Daytona prototype engine, they're benefiting from, but they may not be aware of. We also can't discount the role of the enthusiast. Even for us, Jordan, you know how it works; people find out that we're in the automotive industry, and when it's time for them to

go look for a car or anything automotive related, who do they come to ask? Enthusiasts are opinion-makers. So, although a consumer who isn't an enthusiast may not be aware of what's being done, the enthusiast is very well aware and is influenced by it. And our enthusiasts are heavily influenced by our performance (see Figures 10.12 and 10.13). When I look at motorsports there are so many reasons why we participate. It's obviously in our heritage and very much a family tradition. We participate because the engineering tools are available and provide opportunity to train our engineers. To make them

better, returning back into the mainstream. But if you just asked that question one step above, and if you think about our product, there isn't any other product that has a sport around it unless that sport exists for that product.

▷▷
**10.12** Ford GT, plan view photo featuring radical aerodynamic channels
SOURCE: IMAGE CREDIT, FORD.

▷
**10.13** Ford GT, front three-quarter view photo
SOURCE: IMAGE CREDIT, FORD.

Obviously there are tennis rackets and baseball gloves but those are products to serve the sport. This is a sport to serve our product!

So when we ask ourselves the question, why? Of course, there are so many answers why, but you also have to ask yourself, how could we not? We're in such a unique position of being blessed to work on a product with so much emotional appeal to people that hundreds of thousands, if not millions, will watch the finest executions of that product race against each other. And people are passionate about it, you couldn't ask for anything more! There is certainly an aspect of motorsports that has always been about the driver and the fans cheering for their favorite driver more than the car to a certain extent. It's trended even more so that way in Formula One and possibly NASCAR of late, but it is certainly not the case with endurance racing. That is very much about the manufacturer. And it's a manufacturer and a product that influence people in their day-to-day lives. I don't think very many people will root for an individual tennis player or team because they happen to be wearing Nike, or Adidas … They are Roger Federer fans and whatever he might be wearing may or may not influence them, but that's not what they're cheering for. But in our sport, they actually cheer for Ford!

**Question 4: Looking toward the future, while considering zero emissions and autonomous technology, what are the biggest challenges facing designers in the vehicle development and engineering process and what is your advice to designers and engineers to conquer those challenges?**

▶ Starting off with zero emissions and particularly something like a dedicated **battery electric vehicle**, it's about us as designers and engineers having to really challenge our preconceptions. And you can see how hard this is. For example, Tesla taking three generations before they can get rid of the grill. So challenging not only our preconceptions, but also trying to get a feel for the marketplace's willingness for change. And because so much of the development of the automobile has been around developing the packaging of an internal combustion engine, when you take away the constraints, it can be hard. A real challenge for us is to think about **vehicle architecture**, proportion and shapes, and what they could be if you didn't have to have a radiator, or you didn't have to package the exhaust system, etc. It's certainly not just the aesthetic aspects either. I was literally in a meeting today with a team showing me some project work on a floor that still led up to a tunnel. And I asked what the tunnel was for. And they said, "To package wiring …" Guys, do we really need that much for wiring and why does it have to go in one spot? It may be a conceptual legacy for some engineers because they're often designing around existing platforms and there's a thought that if you don't have to change it, then they won't. But I also think it's sometimes

harder for the designers because, if you started from scratch, you may be so shocking that the consumer just isn't willing to accept it. Perhaps they can't see personal mobility being fulfilled by something that looks like that. I think for **autonomous** vehicles, it's potentially even harder because again we have 100+ years of history designing machines that need to be controlled by humans. Humans have to see and control the direction of the front wheels, the speed of the vehicle, etc. And now, if none of that is required, the obvious questions come up, like should we have all the passengers facing each other? And other issues like how important are the windows for visibility to keep people from getting motion-sick? And what do ingress and egress mean when you remove the constraints of moving around the steering wheel?

Or, what if you don't need a center console because there's no longer a gearshift? What can you do with that space? All these aspects we've trained ourselves as constraints and rules don't need to be there now. And obviously we also have to consider the regulatory aspects that have always presumed that there would be a driver. And even if we scrap the rulebook, and get rid of all of our preconceptions, we as engineers and designers have a lot of work ahead to start thinking clean sheet. But we also have to get a feeling and understanding of how far consumers are willing to go. And that is always a fine balance, as you know.

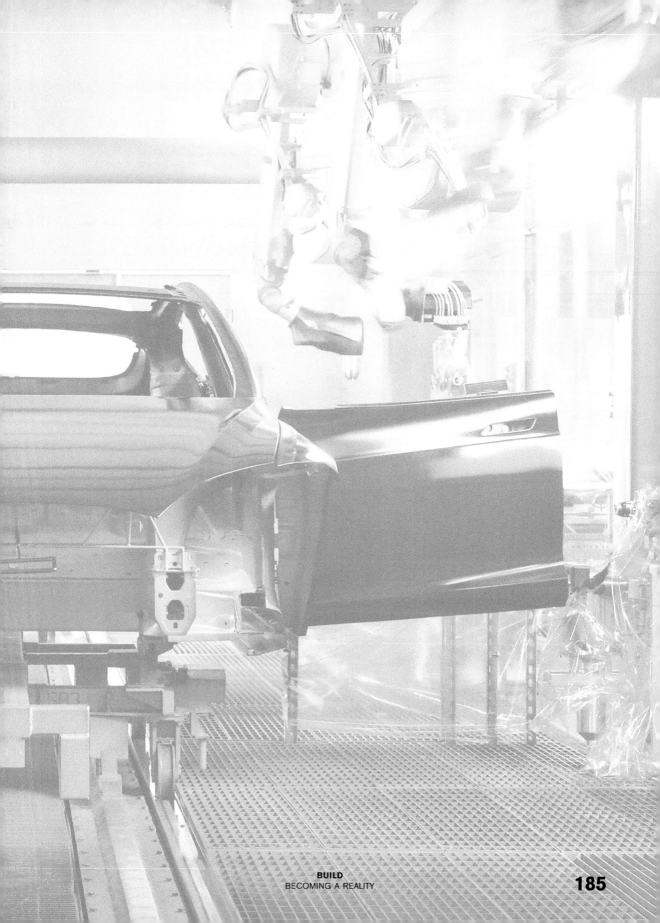

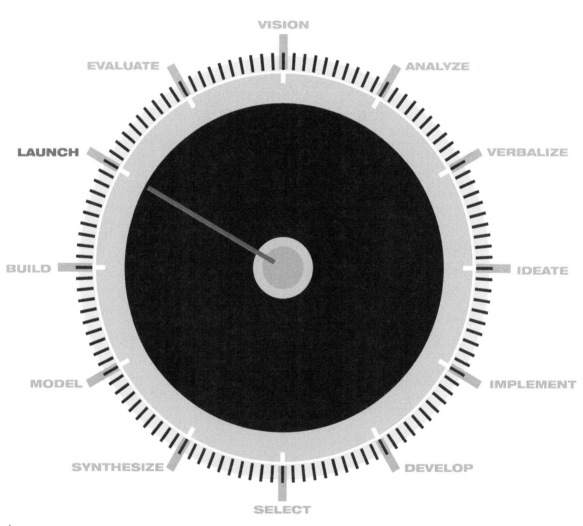

△
**11.0** Process locator gauge
SOURCE: IMAGE CREDIT, JORDAN MEADOWS.

# CHAPTER 11

# LAUNCH

Power to the People:
Telling the Story, Presenting to Management,
the Public, and the Consumer

- ▸ **CRITIQUES,** CONSTRUCTIVE PRESENTATIONS, AND THE EXCHANGE OF IDEAS

- ▸ **PRESENTING** TO CLIENTS, MANAGEMENT, AND KEY STAKEHOLDERS

- ▸ **PITCHING** TO PROSPECTIVE USERS, SELLING NEW VIEWERS ON AN IDEA

- ▸ **LAUNCHING** A VEHICLE

- ▸ **Q&A** MORAY CALLUM, VICE PRESIDENT OF DESIGN, FORD MOTOR COMPANY

In Chapter 10 we covered the importance of vetting an idea, testing, and **market research**. Though this may not be seen as the core of the creative process, it can have an enormous impact on the final design and how it is experienced by the user. In fact, the development and implementation of any **aesthetic principle** should take account of feasibility, manufacturing, craftsmanship, and execution. And to have an understanding of this is to be mindful of the input and contribution of the broader development team.

This chapter will further build on this idea, and explore the importance of communication. In spite of how emotional and personally engrossing the creative process may be, design is very much a team sport. And good communication is essential for any team to function successfully. Added to this, the communication of an idea to the broader public and consumers can be crucially important to launching a vehicle (see Figure 11.0). One never gets a second chance at a first impression. It is crucial to communicate the key aspects of the design effectively, giving

potential users the opportunity to appreciate what went into the vehicle. This is a hugely important point in the creative journey as well. It's the moment that all of the work which has gone before is finally shared with the public. For obvious reasons this can be emotionally charged and often downright scary. However, if the designer has given due diligence to the creative process and maximizes every step along the way, coming up with a strong presentation and communicating the product should be easy. In any case, one has to face the fear, and LAUNCH!

Not all presentations are the same. Having a good feel for the different types is part of being a successful designer. This chapter will provide a brief overview of the typical ones that you might encounter, and how to organize them effectively. The three most common types involve the following: critiques and internal creative discussions, in which case, you're effectively preaching to your choir, although teaching them to sing a new song. It's very much about how you relate to your colleagues. The second most common is really about presenting to the semi-initiated. People who don't know specifically about your proposal but may not be completely clueless either. This is really about transferring key messages. The third type has to do with pitching to potential clients, customers, and users. The main objective here is not only to communicate key messages but also to sell new viewers on your specific idea. Though they have different objectives and outcomes, all three are built around the idea of effectively communicating and telling the story of your design.

## Critiques, Constructive Presentations, and the Exchange of Ideas

Any design is a result of a series of decisions. Decisions are the by-product of discussions and dialogue. In spite of what the media and press would have you believe, vehicle design in a professional context is very much a team sport, much more akin to soccer than a solo endeavor like golf or tennis. An individual may score, but it takes a team to win. Not only is the process quite lengthy, vehicles at the same time can be complex, mandating collaborative teams. It's essential to work effectively together and garner various opinions when

making an important decision on how to move the product forward. Productive critiquing with peers is at the core of the creative dialogue (see Figure 11.1). In short, you're not on this journey alone and having strong allies who can assist and collaborate is the key to success. When relying on others during a critique, it helps to keep a few things in mind to ensure that the conversation is productive. Honesty is crucial. Teams function optimally when there is trust, and trust can never be developed without honesty. Avoid playing favorites and always strive for fairness. Because design can be an emotional process, it's nearly impossible to evolve a design without forming relationships. Almost all productive relationships are built on mutual respect and admiration, and for this reason it's key to follow the Golden Rule; do to your colleagues as you would have them do to you. Finally, it helps to balance one's personal passion by always remaining professional. At the end of the day, never take criticism personally. You should in fact welcome criticism; it's how you grow!

When preparing for a critique and organizing the content to be reviewed, it helps to structure the information into three phases corresponding to a beginning, middle, and an end. Bear in mind that most critiques are to review work in progress, so having a sense of the workflow can give the evaluator a good deal of context. Begin with the goals and objectives. Adding your personal thoughts and a bit of background helps to set the stage. Then, quickly get into showing the current state of affairs. Explain the work that you are engaged in and offer it up for opinions and points of view. Finally, close with conclusions on what has been completed so far, then quickly follow up with challenges to be faced moving forward.

Critiques are a two-way street. When participating as an evaluator, it is important to keep a few things in mind as well. On the flip side, always strive to empathize with your peers, be a good listener and aim to assist. Understand that this is a human process; it is important to be sensitive to individual strengths and weaknesses. We all have our strong points and things we need to work on. Respectively it also helps to understand the unique

△
**11.1** Chrysler 300 design team critiquing and evaluating clay model proposal
SOURCE: IMAGE CREDIT, FCA DESIGN.

challenges of a project. With this in mind, attempt to offer objective measurable items to be improved, with steps on how you would achieve the enhancement. And finally, when asked for an opinion, always give an honest one (see Figures 11.2 and 11.3).

Critiques and constructive discussions are essential for any design project to move forward. Engage in them regularly with partners that you trust to add value. They are an enriching and informative exercise for both the designer and evaluator.

## Presenting to Clients, Management, and Key Stakeholders

Another type of common presentation involves reviewing and sharing ideas with clients, management and key stakeholders. In much the same way a design project cannot proceed without an in-depth understanding of the user, these undeniably important presentations cannot be formed without an understanding of the customer. In this case, the term refers to the client, management, or key stakeholders. As designers, we provide a service to an industrial enterprise. Get to know and understand as much as possible about who is sanctioning your effort. This will not only inform your work but how you

present it. The personalities and unique individual preferences of your customer should never be underestimated. Design can quite often be a subjective exercise so expect that personal taste will weigh in. To that end, ethnic and cultural biases are also very important to consider. For example, a good deal of study has concluded Asian organizational structures are composed quite differently than European ones. As one would imagine, a design presentation for a luxury sedan in a company like Lexus may be orchestrated very differently than in a company like Mercedes, even though the products may have a good deal of similarity. Finally, as with all forms of show business, good timing is essential. A presentation that is too short quite

△△ **11.2** Ford Design team members evaluating and critiquing student projects at ArtCenter College of Design 1
SOURCE: IMAGE CREDIT, ARTCENTER COLLEGE OF DESIGN.

△ **11.3** Ford Design team members evaluating and critiquing student projects at ArtCenter College of Design 2
SOURCE: IMAGE CREDIT, ARTCENTER COLLEGE OF DESIGN.

often won't do the work justice, whereas one that is too long will make even the most exciting subject seem boring.

When presenting to clients, management, and key stakeholders, designers are ultimately looking to move the proposal forward in a productive manner. Your customer may not be design savvy or understand aesthetic principles; after all, this is why they hired you. To maintain the respect, credibility, and trustworthiness, it is helpful to show qualitative and quantitative measurements of success. This will ultimately keep the presentation focused on sharing information, and gaining alignment.

As with almost all presentations, it helps to structure the information in a flow from beginning to middle, to end. Starting with a recap of the business objectives and the specific **design brief** can remind the viewer of the goals. In this regard, it helps to orient the viewer/audience in the overall timeline of the project. This can help to give context and efficiency to the discussion. Then briefly explain the methodologies and design work you're pursuing, relating specifically to that point of the process. For example, early on, the focus would be on research and idea generation, whereas later on in the process, the focus would be on detailed model-making. Finally, in closing, explain how the project is unique and how you are exploring innovative solutions to meet the challenges. With respect to this, always save room for dessert. Everyone loves a pleasant surprise. A small, charming, charismatic addition can almost always lighten the mood, making

the presentation even more enjoyable for the viewer.

**Pitching to Prospective Users, Selling New Viewers on an Idea**
The third most common type of presentation for a designer to master involves capturing the imaginations of new clients and customers, prospective users, and completely uninitiated viewers of the idea. In a sense, this is good old-fashioned salesmanship at its best. This may not come naturally to all. In reality, the act of presenting your design work to someone and convincing them that it has merit not only can be challenging but also downright scary. However, few can argue with the fact that whether it's mastering a kick-starter campaign, an elevator pitch to a would-be client, or presenting a vehicle at a global motor show to the international press, there is great value in being able to present a design effectively.

Although this type of presentation may seem the most daunting, it's actually the easiest. This is because the structure of the presentation is derived from the narrative that you established early on in the conceptual development phase of your project. Furthermore, everything about this type of presentation should ideally relate to storytelling. It has been proven time and time again that the way to generate interest in any idea is to align it with a story. And all of the ideas explored in Chapter 2 of this book also apply when pitching or selling new viewers on your idea. In fact, everything about your pitch can be derived from the basic three-part structures used in almost all storytelling.

*The Three-Act Structure*
For example, step one: your user as you defined them experiences a want or need. Step two: they are confronted with an obstacle or difficulty in fulfilling that want or need. Step three: your product is introduced to them and gives them satisfaction. This is a very common three-phase structure that is composed around the idea of: (1) separation; (2) transition; and (3) reintegration. Simply use the beginning, middle, and an end three-part pattern known as a Three-Act Structure.

- Act I: Set up, most importantly *why* the unmet goal and need of the user.
- Act II: Confrontation, *what* your product is and the design execution.
- Act III: Resolution, *how* it's differentiated and meaningful.

When developing the presentation, there are also some things to bear in mind in terms of how to best deliver the story. The **narrative** should always align with your goal. This is commonly referred to as "staying on message," so be careful not to digress into subplots or examples that are not relevant. Furthermore, the protagonist or hero you are talking about should have some connection with the audience you are presenting to. This is how empathy is established, making it easier for you to hold their interest. Also, be conscious of timing. Most basic narratives can truly be communicated in two or three minutes. To this point, the narrative does not have to run the entire length of your presentation. It's important to not come across as long-winded, or boring the listener with unnecessary detail or information. This is particularly important when pitching a new idea to a prospective client or anyone who might be short on time or attention. Always attempt to be succinct, nailing your point and then moving on. Refine and reduce so only the important details that help paint a picture give definition and clarity to the visuals. And, finally, relax and be comfortable. Rehearse the presentation as many times as needed to become completely at ease with delivering the information. The viewer will respond to your confidence as well as the content being communicated.

Just as the basic structure of your presentation can be broken down into three parts (Act I, Act II and Act III), there are three main components to the content you are communicating. As a designer you should ideally attempt to triangulate the impact of your pitch with three mediums, appealing to sound, vision, and touch. This also helps to captivate your viewers' attention and immerse them in the idea. On the visual front, gorgeous two-dimensional artwork always elicits interest. Even in the age of sophisticated computer graphics, drawn images of vehicles still have a warm character that gives any presentation a human touch. This can also be combined well with complementary virtual data and **3-D** animations. Animations are ideal in presenting technical functions and showing ideal scenarios. At times it's also appropriate to add music to an animation. In these situations, be sure to select music that is complementary to the design work and somehow relevant to the narrative theme. For instance, if you were presenting a sports car for a rebellious personality, rock 'n' roll or hip-hop may be completely appropriate.

Always attempt to provide a tangible three-dimensional object, ideally something that can be touched and appeal to the haptic senses. For example, when presenting an interior idea, even having a sample of an intended material that a person can feel is a plus. Ultimately, models and prototypes dominate any presentation and will always concentrate the discussion. Even having a small 1:18 **scale model** of an exterior car in a boardroom presentation will anchor the room's attention to the designer. Finally, on the auditory front, craft a succinct speech that communicates the narrative and verbally tells how a design is differentiated, meaningful, and appealing to the user. However, if the model and visuals are doing their job, quite often very little needs to be said or verbally presented. Always strive to let the 2-D and 3-D design work speak for itself. Remember that a massive amount of human communication is actually non verbal. In reality, one can often presume that you may not always be present to communicate the key aspects of your work. It's generally better to be biased toward visual storytelling that can be understood by any audience, speaking any language that may not have the ability to hear a verbal explanation. A good speech or verbal description should merely help the presentation along and humanize the experience for the viewer.

Having compelling images, beautifully crafted objects, and a well-honed verbal description are the components of a great presentation. One merely needs to present them in a three-part narrative that communicates *why*

it is meaningful for a user. Doing this effectively will captivate the attention and interest of any prospective client or customer. These basic fundamentals ensure success in academic forums, in any meeting or boardroom, or when presenting a vehicle to the public.

**Launching a Vehicle**
The design process is a journey, has been said several times throughout this book. It's a quest to discover new ways to deliver a unique and meaningful experience to a user. As designers, we seek to use **aesthetic principles** to alter hearts and minds, impacting lives in a positive way. Arriving at the stage where one can share an idea with the public is a special point in that journey. In a sense it is a form of completion and closure. It represents the end of intense internal development, and the beginning of an external dialogue between the user and creator. It is always an emotional moment when the designer first witnesses a vehicle they helped to create being used, viewed, and experienced by others. Hopefully the wants and needs of a user have been captured in a compelling shape. The visual semantics of the vehicle should ideally tell a story. And that story should be appreciated and understood by all who view it. It's a very personal endeavor creating character in vehicles. And as design and technology evolve, the landscape for **transportation design** is shifting. That shift is covered in the final chapter of this book.

# Q&A

**MORAY CALLUM**

VICE PRESIDENT OF DESIGN,
FORD MOTOR COMPANY

*Moray Callum is Vice President, Design, Ford Motor Company (Figure 11.4). Callum leads the design of all concept and production vehicles for the Ford and Lincoln brands globally. Since 2006, as Executive Director, Design, The Americas, Callum has had overall responsibility for the design of all cars and trucks created in Ford's North and South America studios and the new Lincoln products. From 2001 to 2006, he led the design transformation for Mazda, based in Japan. Callum joined Ford in 1995, in the Ghia SpA studio in Italy, after having worked for Chrysler Corporation, UK, and PSA Peugeot Citroën in France.*

△
**11.4** Moray Callum, Vice President of Design, Ford Motor Company
SOURCE: IMAGE CREDIT, FORD DESIGN.

**Question 1: Who are your personal design heroes? Who do you most admire, and why? How does this key influence relate to the work you do for Ford Motor Company?**

▶ I have a few. To go back in time, Flaminio Bertoni, the designer of the Citroen DS. I think the great thing about the Citroen DS is that it's a family sedan effectively, but it is also an extremely exciting-looking vehicle. Not a supercar, not a sports car, it's a practical family car but it's

still a rolling sculpture on wheels that I think is absolutely beautiful and so avant-garde for its day. That is what makes it just amazing. I don't know if he did much else after that, but the DS was enough, to be honest.

Another one on my list is Alec Issigonis, the Mini designer. It's a brilliant piece of engineering that resulted in a brilliant piece of design work. The whole premise behind the car is great: it was one man who created everything from an engineering point of view and the design just came from it.

I also have a few more obscure ones, like David Bache who designed the first Range Rover. That car completely created Range Rover as a brand, but it also was extremely functional at the same time. Now to think of form following function, on a car in a design language that is still viable today, was really impressive.

And then I have a few more logical ones like Guigaro. I think anyone my age admired him growing up, in terms of what he brought to car design and how he changed things. For example, how the first Golf changed expectations of design moving from being decorative to more functional but still beautiful. And Gordon Burig, who designed the Cord and did great work throughout his career.

In terms of impact, I wish they could influence me more, to be honest. They had a much more significant impact on the overall project than we can today. They oversaw the project in a lot of ways both in terms of the engineering and the design. The business is too big now for a one-man-band, so I'm jealous of them in that regard. They weren't just stylists; they needed to understand engineering, practicality, and function to be able to create the products that they created. And I suppose, if anything, I still see myself as reasonably pragmatic when we're developing a design.

Stepping outside of automotive, I would call out Le Corbusier. I think his work is stunning and spans across a broad range of products and design languages.

Though he's known for very rectilinear designs, if you look at the Chapelle Notre-Dame-du-Haut de Ronchamp, it's a lovely mixture of fluidity and rectilinear design. I also love the minimalist designs he did, though not sure I agree with all of them—like the idea of everyone living in tower blocks. His creativity and sense of proportion were superb. He even designed houses around the importance of cars, something he understood the importance of early on, and ahead of many others. I used to live in a town just outside of Paris where there was a house of his called the Villa Savoye. If I'm not mistaken, the door handles and kitchen details were from Voisin! So there was a nice relationship between him and cars.

## Question 2: How important is teamwork to the vehicle design process? How do you foster and encourage collaboration and communication among your designers?

▶ I think that's a really important question, actually. Although teamwork is very important, I think design is at its purest when it's actually a single vision. I think what you need to be able to do is be the caretaker for it throughout the project. To do this, you need to be able to create teams that agree with that vision, and are aligned with it. So in a way, it's about how you take a single point of view and have it become a team's vision. So it may start off as one person designing it, but you then have to get collective ownership.

A part of any leader's job is to get alignment. You need to curate that alignment and see that it's consistent. The objective is to get the finished result the way you want it to be. The other important thing to do is to ensure there is always a safe space to come up with new ideas, and allow enough fluidity to keep refining and editing those ideas within a broader vision. We work in a very processed, disciplined, sometimes digital world; we try to allow that element of creativity that is not black and white to flourish. And sometimes it's part of how you empower the team, getting everyone aside from them to understand the creative process. Sometimes there isn't a right answer, or wrong answer, there's just an answer and it needs to be developed.

## Question 3: What types of strategies and techniques do you rely on when presenting new vehicle design proposals to senior management and key stakeholders?

▶ I always like to think that a design should speak for itself. Hopefully you don't have to explain it too much or rationalize why you've done things a certain way. If you've done the job right, you can usually present it by showing the thing and just standing back. I think sometimes people expect us to justify what we're doing, but if a design is right, you don't need an explanation of why it's right. You shouldn't need a rationale. It's a bit like the discussion we just had with the last question; you can't rationalize it by a set of yes or no questions. For me, it's one of those areas where hopefully the beauty is in the eye of the beholder and you're actually getting their confidence just by showing them the design (see Figure 11.5).

Mustang's a great example of that; we didn't need to explain why we were doing things on the

△
**11.5** Ford Performance Vehicles
SOURCE: IMAGE CREDIT, FORD DESIGN.

car, people just got it. The GT is another great example. The explanation is actually superfluous to whether a person likes the vehicle or not.

**Question 4: Looking toward the future while considering zero emissions and autonomous technology, what are the biggest challenges facing designers in communicating and presenting new vehicle solutions? What is your advice to designers to conquer those challenges?**

▶ I think there are challenges and pitfalls in designing products like these because there is an expectation to look completely different. For example, if you look at the Tesla Model S, that people quite like, it doesn't look like a science project. It looks like a pretty car. I think the standards for how people will judge aesthetics won't change. There's a tendency to go off and do something a little bit more different or maybe sometimes even wacky. Hopefully, in the future, there will be opportunities with packaging and proportions so we can make these cars look better as opposed to just different. I would warn designers away from doing different for the sake of different.

Now, we have a tendency to talk about exteriors, but the future holds a bigger challenge for interior design, how to make the technology user-friendly, and how to limit the confusion for the user. There is so much more technology coming. Design definitely has to go through an editing process to say what's important and what's not, and how to visually communicate that new technology in an intuitive way.

**194**

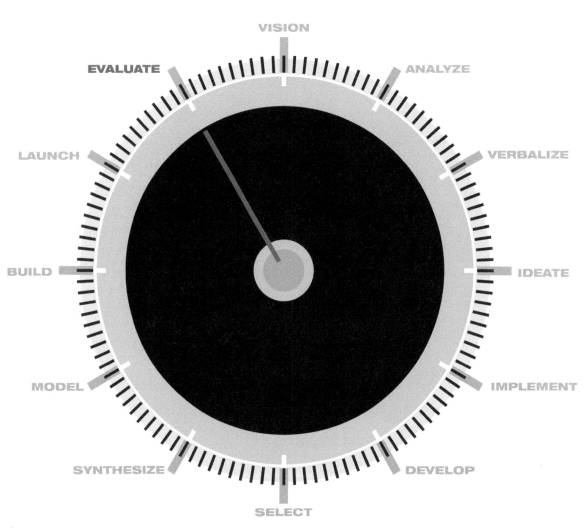

VISION

ANALYZE

EVALUATE

VERBALIZE

LAUNCH

BUILD

IDEATE

MODEL

IMPLEMENT

SYNTHESIZE

DEVELOP

SELECT

△
**12.0** Process locator gauge
SOURCE: IMAGE CREDIT, JORDAN MEADOWS.

# CHAPTER 12

# EVALUATE
Taking Stock of Lessons Learned and
Looking Toward the Continuous Improvement
of the Idea and Its Replacement

## Taking Stock of Lessons Learned while Looking Ahead

Early in this book, we took a deep dive into the importance of answering the *why* component of what compels a user toward a particular vehicle. The aesthetic draw and the principles that compose its visual semantics are at the very core of that attraction. In presenting a vehicle to the world, telling that **narrative** is key to getting this point across. As stated before, this is a pivotal moment when an idea transfers from an internal exercise, pursued privately by professionals, to an object that is shared and appreciated in public. From that point on, a special type of dialogue takes place by which users can form, shape, and impact decisions made on successive vehicles. The bottom line is design is an ongoing process that is cyclical in nature. Each decision that a designer makes and is experienced by a person becomes fuel and fodder for successive decisions and future products (see Figure 12.0).

The creative process is analogous to a hero's journey, in which a designer delves deep into the unknown to discover solutions that are applicable to others. In releasing a product, the designer reaches the moment in their journey when they've returned from the quest to share the newfound wisdom. This wisdom propels future journeys and serves as a catalyst for the entire cycle to begin again.

One must understand where you've been to know where you're going in the future. Betterment and continuous improvement are one of our collective goals in life. For this reason, it is crucial to understand the lessons learned. Honest analysis of the strengths and weaknesses of a particular vehicle is crucial to producing a replacement that is better. Every professional designer immediately knows what they would do differently once they see a vehicle being used in public. It's almost a reflex. Now more than ever, this questioning of how to go about things differently in the future is taking place in the profession on a broad scale.

## Looking Forward: Factors Changing the Future of Transportation

At the time of writing this book, the **transportation design** industry is undergoing massive

change. Macro technological and societal trends are shifting nearly every assumption of how vehicle manufacturers will maintain and generate profits moving forward. Various researchers in the industry believe there are five key factors that will affect the way we move through our lives in the future. These five include population growth and shifting demographics. In almost every projection of the future, the world will be dealing with many more people than it has now. The second of five includes new mobility paradigms. Due to the socioeconomic conditions in both emerging and established mobility markets, people are choosing new ways to move about. The third factor includes ubiquitous connectivity and the Internet of Things. Virtually every powered object in our lives may be potentially linked in some way online. The fourth significant factor involves **zero emissions powertrains** and the proliferation of battery technologies. The fifth factor, and potentially most pivotal, involves **autonomous technology** and intelligent self-driving vehicles.[1]

All these key factors are well underway and will have an effect on the traditional ways in which vehicle manufacturers operate. The question is, what will be the scale and magnitude of change? How, when, and who will be able to adapt? And what types of emerging challenges will designers face in this dynamic and shifting new world?

**Future Population Scenarios**
Population growth is an undeniable part of our future! No one can accurately predict the future. In many cases, accurately predicting long-term scenarios for any business is a bit like forecasting the weather. However, there are

△
**12.1** Composite image of future traffic scenario
SOURCE: IMAGE CREDIT, JORDAN MEADOWS.

△△
**12.2** Composite image of future city scenario 1
SOURCE: IMAGE CREDIT, JORDAN MEADOWS.

△
**12.3** Composite image of future city scenario 2
SOURCE: IMAGE CREDIT, JORDAN MEADOWS.

some basic assumptions that are likely to evolve from our current reality. You would be hard-pressed to find an expert who doesn't agree that population growth will be something that all of us will face. Imagine that anything you enjoy doing presently will definitely be shared by many more people as you get older. And although demographers predict significant

growth in emerging markets, such as India and China, everyone in the world will be faced with the global impact of significantly more people on planet Earth[2] (see Figures 12.1–12.3).

Most of the growth will take place in cities around the world. This is due to the fact that cities already have fairly large populations that

Jordan Meadows Design

△ **12.4** Spectrum of mobility and transportation applications
SOURCE: IMAGE CREDIT, JORDAN MEADOWS.

will regenerate. They will also increase their population, scale, and density, drawing people to them seeking jobs. However, there are different types of cities and their unique dynamics will drastically affect vehicle usage. For example, cities with large populations distributed across a large sprawling area such as Los Angeles will require very different solutions to cities that are composed of a large population in a dense and compact area. Furthermore, the per capita spending power and wealth in some cities versus less affluent developing areas will result in different types of vehicles and usage patterns. Traditionally, vehicle manufacturers attempted to create one type of vehicle for all cities. Moving forward, this may not be the case. Even with regard to public transportation, new or emerging cities with a larger landmass will have the opportunity to pursue different types of solutions versus older, more established urban centers. When laying out new roads in a growing area, it may be possible to establish dedicated tram lines to avoid traffic, whereas an older dense area may find this challenging.

## New and Emergent Mobility Paradigms

The proliferation of ridesharing services such as Uber and Lyft are clearly disruptive to the taxi industry as a whole. One could argue these services not only may change the way vehicles are used, but could potentially change the

way vehicles are manufactured and sold as well. Not only are rides available on demand to nearly anyone anywhere, retail is also available on demand. The proliferation of services such as Amazon, Insta-cart, and other retail delivery providers, is fundamentally shifting the need and frequency of visits to brick-and-mortar stores. Added to this, there is a general shift away from traditional ownership models. The rise of the "**sharing economy**" across all goods and services cannot be ignored. There's been a clear and definite shift away from the assumption that the consumer needs to own something to experience it.[3]

Conversely, if one does own something, one can also potentially benefit financially by sharing. In a sense, services such as Airbnb have not only impacted traditional business models such as the hospitality industry, they've also altered many consumer's perception that they actually need to go out and buy anything. This is true, whether it's record sales for hard copies and recordings, or vehicles to a certain extent. There is also a generational dynamic to these new mobility paradigms. Younger buyers are not only buying fewer cars, they're also much slower to apply for and obtain a driver's license than generations before. What used to be a sign of independence and freedom for older generations simply does not have the same cachet anymore.

Not only are users rethinking traditional ownership and usage models, when they do choose to purchase a transportation experience, they are provided with an increasing array of options. This can range from public bike rentals in urban centers, to micro-vehicles provided by services like Zipcar, to new commuter rail services. It's not hard to imagine that in cities of the future shared commuting helicopter services may one day be likely (see Figure 12.4).

## Connectivity and the Internet of Things

One can hardly remember what life was like before the age of the internet. In reality, just the mere idea of life without a cell phone now seems very difficult to imagine. This dependence on connectivity is here to stay and will begin to extend to nearly all of the powered products that surround us. Connectivity will not be relegated to just information and communication devices such as televisions, computers, and phones. It will eventually extend to electronics, appliances, and any form of machinery that one can imagine. This will effectively allow users to access and control these objects in completely new and profound ways. It will also potentially change these objects into data collection devices[4] (see Figure 12.5).

Because vehicles are an assemblage of numerous products, they occupy a unique place in the spectrum of

EVALUATE
TAKING STOCK OF LESSONS LEARNED

△
**12.5** Connectivity between objects, applications, and transportation modes
SOURCE: IMAGE CREDIT, JORDAN MEADOWS.

interconnected objects. They are mobile, transporting us, and offer endless possibilities for connectivity. They move us through our world and provide us with the ability to access and share experiences. Because most of us are completely reliant on some form of vehicle, they seamlessly weave innovation, design, and technology into our lives. And because where we go and how we get there say so much about who we are, vehicles have the potential to become the ultimate data collection device.

Privacy, who owns this data, what they do with it, and the implications of their agenda are a controversial subject of a broad-based ethical discussion. This ongoing conversation is not immediately related to the **aesthetic principles** of vehicle design; however, every product designer should be aware of the sociocultural impact of the objects they create. And now more than ever, these objects

are interconnected in new and complex ways (see Figures 12.6–12.9).

### Zero Emissions Technology

If connectivity and the internet have radically changed the way we share information and data, zero emissions technology has the potential for a similar impact on the architecture of vehicles. At present, the two main areas for investigation involve hydrogen fuel powering combustion engines, and battery-powered systems propelling electric motors. The latter has the most radical benefits to the architectural configuration of conventional automobiles as it eliminates the combustion engine; the heaviest, most technically complex part of the **vehicle engineering** equation. In their most efficient configuration, **battery electric vehicles (BEVs)** use a skateboard-type frame where the cells are packaged below the vehicle floor, in the area between the wheels. Electric

motors can then be packaged at either axle, within the diameter of the wheels. This frees up an enormous amount of space in the vehicle, allowing for seemingly endless flexibility for packaging above these elements. This allows designers to explore radically new **silhouettes** and fundamental shifts to the architectural composition of vehicle bodies (see Figures 12.10 and 12.11).

The benefits aren't limited to only vehicle design and architecture. Consider that when combined with solar power systems, they can have a very low impact on the environment. Battery technology not only allows for a more efficient usage of space within a vehicle, but can also provide a way to use cleaner energy. This has the potential to radically impact nearly every socioeconomic and political system related to fossil fuels.

Due to shifting economies of scale, battery technology was previously considered premium.

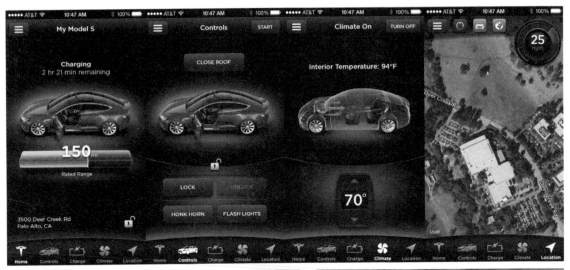

△△△
**12.6** Connectivity between automobile functionality and smart phone applications 1
SOURCE: IMAGE CREDIT, TESLA.

△△
**12.7** Connectivity between automobile functionality and smart phone applications 2
SOURCE: IMAGE CREDIT, FORD MOTOR COMPANY.

**12.8** Ford Mustang interior with Sync 3 infotainment system
SOURCE: IMAGE CREDIT, FORD MOTOR COMPANY.

△
**12.9** Ford Sync image showing connectivity between automobile functionality and smart phone application
SOURCE: IMAGE CREDIT, FORD MOTOR COMPANY.

△△
**12.10** Tesla Model-S, key components featuring battery pack located at the bottom of image, displaying flexibility of upper body design
SOURCE: IMAGE CREDIT, TESLA.

△
**12.11** Tesla Model-S, key battery and technical components, side view. Many electric vehicles rely on the similar packaging to provide maximum freedom in body shape and design.
SOURCE: IMAGE CREDIT, TESLA.

**EVALUATE**
TAKING STOCK OF LESSONS LEARNED

**203**

△ △
**12.12** Tesla Giga factory, a key
enabler for providing battery
technology at a mass scale
SOURCE: IMAGE CREDIT, TESLA.

△
**12.13** Tesla Model-S, electric motor
and battery pack components
SOURCE: IMAGE CREDIT, TESLA.

Moving forward, the cost of
electrification per vehicle has
the potential to become more
affordable and proliferate into
mass-market applications. Added
to this is the fact that technology
advances are enabling the
previously limited range for such

vehicles to rise (see Figures 12.12
and 12.13).

Hydrogen-fueled vehicles are
also emissions-free. However,
because the technology is
reliant on a combustion engine
and special safe packaging of

hydrogen tanks, it offers vehicle designers less freedom from an architectural perspective. Both require the development of a convenient refueling, or recharging, network. In any case, many governmental agencies around the globe are mandating that vehicle manufacturers include zero emissions technology as a percentage of their overall vehicles produced. Many offer tax incentives for consumers as well. The adoption of zero emissions powertrains, be they electric or hydrogen, offers new opportunities for designers as well as undeniable benefits for the environment (see Figures 12.14–12.17).

△
**12.14** Tesla Model-S, cutaway view featuring ultra-compact packaging of technical elements allowing for flexible space and added storage areas
SOURCE: IMAGE CREDIT, TESLA.

▷
**12.15** Tesla Home-Charging Station
SOURCE: IMAGE CREDIT, TESLA.

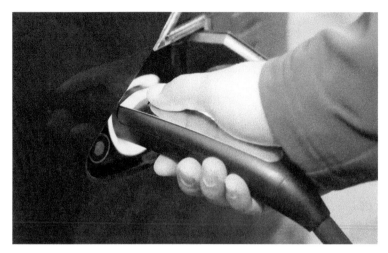

**12.16** Tesla Model-S, charging port
SOURCE: IMAGE CREDIT, TESLA.

**12.17** Tesla Supercharger Station, featuring free connectors that charge their vehicles in minutes rather than hours
SOURCE: IMAGE CREDIT, TESLA.

## Autonomous and Driverless Scenarios

The automotive industry has been on a long and steady campaign to make vehicles safer. Over time, a good deal of effort has gone into managing how vehicles behave when an accident occurs. More recently, a similar amount of effort has been focused on accident avoidance systems (preventing them from happening in the first place). As technology has advanced, and the amount of computing power in the average automobile has grown, these safety-focused programs have evolved into a new generation of advanced driver assistance systems (ADAS) that involve a host of different products from "lane assist" to "automatic braking" systems and "crash avoidance" to "cruise control," etc. Combining ultra-advanced versions of these driver assistance applications has formed the

△
**12.18** Tesla Autopilot, illustration depicting array of vehicle sensors
SOURCE: IMAGE CREDIT, TESLA.

basis for autonomous, self-driving vehicle technology (see Figure 12.18).

Vehicle autonomy, or the self-driving car, may represent one of the most pivotal shifts in mobility since the inception of personal automotive transport. Nearly every aspect of automotive design and engineering for over a century has been centered on the idea of an individual taking active command of the machine. In reality, many might find the basic concept challenging. It's really quite understandable. Being in control, the joy of driving, and a sense of empowerment are massive parts of the emotional appeal of vehicles. Additionally, many people have a natural sense of skepticism and mistrust for

new technology. There are also significant regulatory, legislative, and liability hurdles to overcome. Who would be responsible in an accident with an autonomous vehicle, for example? How would insurance companies deal with such a class of vehicle? With regard to trust in the technology, the actual real-world consumer acceptance can only be truly evaluated gradually. Due to this and many other complex technical factors, the industry will be dealing with a period of semi-autonomy for quite some time. This basically means that vehicles will have the ability to drive themselves; however, human intervention will be required at times. In any case, the paradigm has definitely shifted, and it doesn't take much to imagine

where the technology is headed (see Figure 12.19).

Semi-autonomy will eventually give way to fully autonomous vehicles. Traditional automobiles with steering wheels and pedals that possess some autonomous capability will eventually give rise to vehicles designed to function without any human intervention. In theory, when this point is reached, users will have the opportunity for radically different usage scenarios. Every aspect of our interaction with vehicles could potentially shift. The nature of our relationship with transportation could be altered. Whereas today we're in direct control of vehicles, the future may bring a day when we trust our lives to digital transportation devices with some

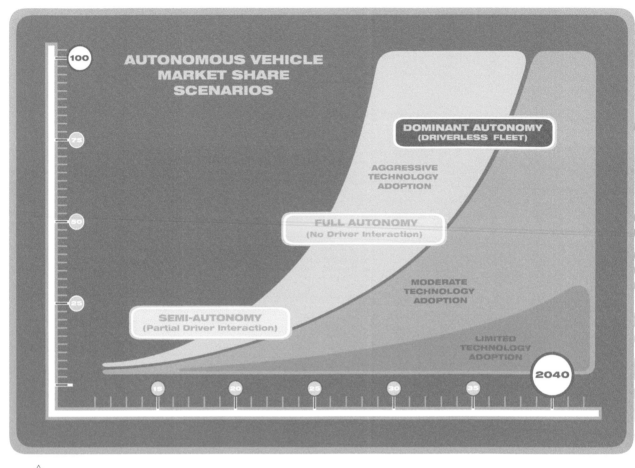

AUTONOMOUS VEHICLE
MARKET SHARE
SCENARIOS

DOMINANT AUTONOMY
(DRIVERLESS FLEET)

AGGRESSIVE
TECHNOLOGY
ADOPTION

FULL AUTONOMY
(No Driver Interaction)

MODERATE
TECHNOLOGY
ADOPTION

SEMI-AUTONOMY
(Partial Driver Interaction)

LIMITED
TECHNOLOGY
ADOPTION

2040

△
**12.19** Autonomous technology possible adoption rates of moving forward in the
future, adapted from McKinsey Automotive and IHS Insights studies
SOURCE: IMAGE CREDIT, JORDAN MEADOWS.

▽
**12.20** Velodyne LIDAR sensor, essential for enabling
autonomous vehicles to read their surroundings
SOURCE: IMAGE CREDIT, FORD.

degree of artificial intelligence.
The change in our relationship to
personal modes of transportation
will undoubtedly alter how we use
our time both inside and outside of
vehicles. It will have an impact on
our communities, cities, roads, and
public spaces. In principle it will
also offer enhanced safety.

As with all new technology,
autonomous self-driving vehicles
offer new possibilities that will
also pose new questions and new
responsibilities. The very nature
of how these vehicles will be
integrated into the fabric of our

EVALUATE
TAKING STOCK OF LESSONS LEARNED

lives will be the result of decisions made by designers moving forward (see Figure 12.20).

## New Questions for Designers

When these five key factors are combined, the industry is faced with new questions. In combining these evolving macro trends, one quickly understands the potential for radical disruption of traditional business models in the world of vehicle design, marketing, and manufacturing. Consider the bygone era when the average person purchased hard copies of music rather than downloading and sharing. The record industry experienced massive change due to technology. Moreover, one could argue the very nature in which we appreciate music shifted. The world of imaging has experienced a fundamental shift as well. Analog cameras were at one time quite popular. Large corporations such as Fuji and Kodak were quite successful in offering and developing film. Digital imaging has undeniably altered the way most of us capture, process, and share images. From journalism to publishing, the list of traditional industries upended by the advent of new technology is not a short one. Anyone actively engaged in the vehicle and transportation sector should consider where they stand in this regard. This is particularly crucial for designers as they are directly responsible for the aesthetic, human, and emotional qualities of the user experience.

These macro trends indicate that the way many of us purchase and use our vehicles is likely to change. While some consumers will always opt for traditional vehicle ownership, one can imagine the day when shared usage will become ubiquitous. As demand for different types of mobility grows, and technology offers ways to make mobility less burdensome, users will have a broader range of ways to access transportation. Many of these choices will not involve going into a traditional dealership every few years to purchase or lease a vehicle.

The key question for traditional vehicle manufacturers then becomes how to monetize mobility. They will have to remain competitive in their current areas of expertise. However, many of the opportunities for growth moving forward will not be from the sources familiar to them. Future growth will come as a result of new business propositions created by the confluence of new technologies and sociocultural trends. To remain competitive and grow, traditional vehicle manufacturers must evolve to include services and experiences.

As the competitive landscape is redefined to include services and experiences in the technology age, new competitors will emerge. Where vehicle manufacturers traditionally competed among themselves, they are now potentially faced with competition from a range of technology-based service providers. For example, the big three at one point in time referred to General Motors, Ford, and Chrysler. It then evolved to more accurately refer to Toyota, Volkswagen, and General Motors. Moving forward, the three key players in the transportation world may not only include traditional automotive manufacturers. As a result, core competencies will be reframed. The scope of product development and design will undoubtedly evolve. To be successful in this fluid and competitive world, successful vehicle manufacturers will have to simultaneously learn new skills and leverage partnerships when necessary. In fact, the central focus of some organizations may shift in a quantum way. Consider if a company like Apple had only remained a computer manufacturer and ignored mobile phones or music. Just imagine then how key players in the vehicle world may involve moving forward.

## Key Implications for Designers

As stated before, creating future scenarios for any business can be a bit like predicting the weather. In considering these key factors and their impact on the world of vehicle design and transportation, it's important to remember that none of them are actually etched in stone. It is true that they exist currently; however, they are evolving to varying degrees and subject to whatever unpredictable twists, turns, and occurrences the future may hold. Change may take place slowly, or it may occur rapidly, but in either case you can be certain that much of the timing and outcome is truly uncertain. This is one of the things that makes design exciting.

We can be certain, however, that people will always have a need for transportation. For as long as we've been around, we've had the need to get around. Furthermore, we've always had a desire to appear a certain way while traveling. Vehicles are and will continue to be an integral part of our society, from the days of horse-drawn wagons to autonomous self-driving automobiles. And considering that horses are intelligent animals and stories do exist of them navigating on their own, one can draw parallels.

Through the ages, history and human nature have also shown us that **narrative** is crucial to the way we view ourselves and how we interact with the world around us. The need for visual meaning is deep within and informs the way we create and appreciate design on a psychological level. It's the framework for the aesthetic principles behind everything we create, see, and experience. This phenomenon will remain vital in the way transportation providers offer and communicate their goods and services in the future. As long as humans have different personalities with opinions on how they would prefer to move through the world, designers will offer an array of choices to suit their needs.

With emergent technology bordering on mysterious and magical, trust and promise become ever more important. Users will need to see and understand that their wants, needs, and safety have been given the utmost importance. Vehicles will need to not only offer transportation, but also communicate a sense of advancement and well-being. And now more than ever, designers are the ones to deliver that message and shape a vision of a better tomorrow!

**Notes**

1   McKinsey & Company. *Automotive Revolution – Perspective Towards 2030: How the Convergence of Disruptive Technology-Driven Trends Could Transform the Auto Industry* (Advanced Industries, 2016). Available at: www.mckinsey. de/files/automotive_revolution_ perspective_towards_2030.pdf

2   United Nations department of economic and social affairs – *World's population increasingly urban with more than half living in urban areas.* Available at: www.un.org/en/development/ desa/news/population/world-urbanization-prospects-2014.html

3   *The Rise of the Sharing Economy* (Thinkers 50, 2013). Available at: thinkers50.com/wp-content/ uploads/The-Rise-of-the-Sharing-Economy.pdf

4   *How will the Internet of Things look by 2025?* Lee Rainie. (Pew Research Center 2016). Available at: www.pewinternet. org/2016/03/29/how-will-the-internet-of-things-look-by-2025/

# ABOUT THE AUTHOR

**JORDAN MEADOWS**
*was a key member of the Mustang design team. Prior to Ford, he served as Design Manager at Mazda. While there, his team completed the noteworthy Nagare concepts. Prior to this, Meadows was a Team Leader at Volkswagen. During his time at Volkswagen, Meadows assisted the realization of an advanced studio near Berlin. Meadows began at Chrysler creating the Jeep Willys and Dodge Kahuna concepts. A graduate of the Royal College of Art, and the Rhode Island School of Design, he now resides in Los Angeles, and serves as a Design Specialist at Ford's Global Advanced Design Group and a faculty member at the ArtCenter College of Design (see Figure 12.21).*

# GLOSSARY

Glossary words appear in bold on their first occurrence in the chapter. Some definitions in the Glossary have been paraphrased from the sources below and blended with the author's personal knowledge:

*Merriam-Webster*
    www.merriam-webster.com/
    dictionary
Wikipedia
    http://en.wikipedia.org/wiki/

**3-D Design (or development programs)** The process of using specialized computer software to develop a digital representation of the surface of any three-dimensional object.

**Aerodynamic** (1) The qualities of an object that affect how easily it is able to move through the air; (2) Designed in a way that reduces wind drag and thereby increasing fuel efficiency.

**Aesthetic (or Aesthetic Principles)** (1) The philosophical theory or set of principles governing the idea of beauty at a given time and place; (2) A set of principles underlying and guiding the work of a particular artist or artistic movement; (3) The philosophy of art; the study of beauty and taste. The term aesthetic is also used to designate a particular style, for example "Japanese aesthetics." Aesthetic design principles include ornamentation, edge delineation, texture, flow, solemnity, symmetry, color, granularity, the interaction of sunlight and shadows, transcendence, and harmony.

**ALMS (American Le Mans Series)** A sports car racing series based in the United States and Canada. It consisted of a series of endurance and sprint races, and was created in the spirit of the 24 Hours of Le Mans.

**Archetypes** See *Carl Jung*.

**Autonomous Technology** Any kind of technology that allows for function without human input; in relation to vehicles, it is a car that is capable of driving, sensing its surroundings, and navigating without human input.

**Autonomy (Autonomous Technology/Vehicles)** The state of functioning, existing, or driving independently.

**Battery Electric Vehicle (BEV)** A vehicle that uses batteries and electric motors instead of an internal combustion engine for propulsion.

**Body on Frame** An automobile construction method; mounting a separate body to a rigid frame that supports the drivetrain was the original method of building automobiles and continues today.

**B-Pillar** The vertical or near vertical supports of a car's window area, designated as A, B, C, or (in larger cars) D. The B-pillars reside where the driver and passenger-side windows end.

**Carl Jung** Psychologist, Carl Gustav Jung used the concept of archetypes in his theory of the human psyche. He believed that universal, mythic characters (see *Archetypes*) reside within the collective unconscious of all people and are the psychic counterpart of instinct.

**Clay Modeling** The process of manipulating clay which can be used to sculpt shapes and figures. As related to vehicle design, it is a way to transform a sketch into a three-dimensional object so it can be studied and reviewed.

**CNC (Computer Numerical Controlled) Machine** A highly automated device that uses computer aided design (CAD) and computer-aided manufacturing (CAM) programs to produce an object by cutting material or building up material in layers.

**Competitive Benchmarking** The continuous process of comparing a firm's practices and performance measures with one of its most successful competitors.

**Concept Car** A car made to showcase new styling and/or new technology. Often they are shown at motor shows to gauge customer reaction to newness and may or may not be mass-produced.

**Cowl Point** The top portion of the front part of an automobile body, supporting the windshield and dashboard.

**Design** (1) Creative problem solving; (2) The creation of a plan or convention for the construction of an object or a system creating a solution to a problem; (3) To plan or make something for a specific use or purpose; (4) To plan and make decisions about how something will be built or created.

**Design Brief** An overview for a design project developed by a business need, the focus is on the desired results of the design, not the aesthetics.

**Design Movement** A specific philosophy or ideal that is followed and promoted by a group of artists for a defined period of time, examples are: Abstract Expressionism, Art Deco, Minimalism, Cubism, etc.

**Design Strategy** A discipline which helps individuals and firms determine what to do, why to do it, and how to inform selections, both immediately and over the long term. The process involves the interplay between design and business strategy.

**Digital Design (or Digital Sketch Modeling)** The process of using a computer and software programs to create pixel-based images and renderings of prototypes and/or graphics.

**DNA (Brand-Concept)** Words, images, and experiences that are the fundamental elements of an organization, creating clarity and unity about its vision and purpose.

**Engineer** A practitioner of engineering, concerned with applying scientific knowledge, mathematics, and ingenuity to develop solutions for technical, social, and commercial problems. An automotive engineer incorporates elements of mechanical, electrical, electronic, software, and safety engineering as applied to the design, manufacture, and operation of motorcycles, automobiles, and trucks.

**Ethanol-Fueled** A vehicle running on 100% ethanol fuel or a mix of ethanol and gasoline (flex-fuel). Ethanol fuel is ethyl alcohol and most often used as a motor fuel, mainly as a biofuel additive for gasoline.

**F1 (Formula One)** The highest class of open-wheeled, single-seat, auto racing that is sanctioned by the FIA (Federation Internationale de l'Automobile).

**Fabricator** A person who is skilled in constructing a complex finished product from a design idea.

**FCA Group** Fiat Chrysler Automobiles, formed in 2014 through the merger of Fiat and Chrysler, it is the seventh-largest automaker in the world that designs, engineers, manufactures, and sells cars, light commercial vehicles, components, and production systems worldwide.

**Hard Model** A non-functioning prototype created to evaluate appearance and design.

**HMI (Human–Machine Interface)** Any design application that presents information to an operator or user with intent to implement control instructions, typically displayed in a graphic format, and often governed by software.

**H-Point** The theoretical relative location of an occupant's hip: specifically, the pivot point between the torso and upper leg portions of the body, as used in vehicle design.

**HVAC (Heating, Ventilation and Air Conditioning)** The technology of vehicular environmental comfort, its goal is to provide thermal comfort and acceptable occupant air quality.

**Kodo Design Language** The appearance and design DNA associated with Mazda Motor Corporation's vehicles, typified by sophisticated, sculptural, but minimalistic surfaces.

**Le Mans** The world's oldest active sports car race in endurance racing, held annually since 1923 near the town of Le Mans, France. A 24-hour race which is a mix of closed public roads and a specialist racing circuit, in which teams have to balance speed with their car's ability to withstand mechanical damage.

**Market Opportunity** A situation in which a company can meet an unsatisfied customer need before its competitors. A newly identified need, want, or demand trend, that a firm can take advantage of because it is not being addressed by the competitors.

**Market Research** An organized way to gather information about specific target markets or customers.

**Market Target (Target Customer)** A particular group of consumers at which a product or service is aimed.

**Mill (or Digital Milling)** The computer-controlled machining process of using rotary cutters to remove material from an object by advancing (or feeding) in a direction at an angle with the axis of the tool. It covers a wide variety of different operations and machines, on scales from small individual parts to large, heavy-duty operations.

**Mission Statement (Corporate Vision)** A statement which is used to communicate the purpose and/or goals of an individual or organization.

**Mobility (Mobility Design)** Referring to the broadest definition of transportation products, experiences, and systems capable of moving people or objects from one place to another.

**Modeler** A person who makes models, especially from a plastic medium such as clay.

**Monocoque** A structural design approach whereby a vehicle's stresses and loads are supported through its body or external skin.

**Muscle Car** Any group of American-made, two-door sports coupes with powerful engines, designed for high performance driving.

**Narrative** The art, technique, or process of telling a cohesive story.

**NURBS (Non-Uniform Rational Basis Spline)** A mathematical model commonly used in computer graphics for generating and representing curves and surfaces. It offers great flexibility and precision for handling both analytic (surfaces defined by common mathematical formulae) and modeled shapes.

**Occupant Sightlines (Vision Angles)** The area available for driver or occupant observation out of a vehicle, not obscured by bodywork, or componentry.

**Performance Car** An automobile that is designed and constructed specifically for driving enthusiasts, performance cars are road vehicles. Specially designed racecars are not regarded as performance cars, but performance cars are often raced.

**Pony Car** An American class of automobile launched and inspired by the Ford Mustang in 1964. The term describes an affordable, compact, highly styled car with a sporty or performance-oriented image.

**Powertrain** The main components that generate power and deliver it to the road surface, water, or air; this includes the engine, transmission, drive shafts, differentials, and the final drive.

**Proportion** The relative size, scale, and relationship of various elements in a design.

**Quadrant Positioning Charts** A visual communication aid divided into four sections that graphs the positioning of products, experiences, or brands based on predetermined attributes.

**Quarter Scale** The percentage of proportional reduction for a design study model.

**R&D (Research and Development)** Activities in connection with corporate innovation situated at the front end of the product development lifecycle. R&D departments develop new products or may do applied research within their field to facilitate future product development and design.

**Rapid Validation Mock-up** A low fidelity simulation, model, or prototype constructed to quickly prove a theory or idea.

**Rapid Visualization** A technique used by graphic artists to create a drawing of a concept in several stages. After the completion of thumbnail drawings, a preferred drawing is selected and rendered in full size, usually in pencil. Then a sheet of paper (layout bond) is put over the drawing and it is redrawn/traced with corrections, additions, and alterations. This process is repeated several times, often with color added at some stage, until the image is perfected to the desire of the artist.

**Rotary Engine** A type of internal combustion engine using an eccentric rotary design to convert pressure into rotating motion, conceived by German engineer, Felix Wankel.

**Scale Model** A representation or copy of an object that is smaller than the actual size of the object being represented, and very often used as a guide to making the object in full size.

**Semiotics** The study of signs and symbols as elements of communicative behavior.

**Sharing Economy** (1) Also known as collaborative consumption, a trending business concept that highlights the ability of individuals to rent or borrow goods rather than buy them; (2) A hybrid market model which refers to peer-to-peer-based sharing of access to goods and services, coordinated through community-based online services.

**Silhouette** The outline or general shape of an object.

**Storyboard** An organized form of illustrations or images displayed in sequence for the purpose of visualizing and defining the creative concept or idea.

**Surface Language** An overarching theme or characteristic to an object's three-dimensional finishing used to express a thought and/or feeling.

**SWOT** An acronym for Strengths, Weaknesses, Opportunities, and Threats: A structured planning method that evaluates those four elements of a project.

**Track Car** A car suited in some way to be used on a racetrack.

**Transportation Design** The creative development of vehicle-based products, experiences, and systems capable of moving people or objects from one place to another.

**Uni-body Construction** Sometimes referred to as monocoque, it blends both the frame and body into a single unit, onto which reinforcements are added in specific zones.

**UX (User Experience) Design** The process of enhancing user satisfaction by improving the usability, accessibility, and pleasure provided in the interaction between the user and the product.

**Vehicle Architecture** The manner in which the components of transportation experiences are organized and integrated.

**Vehicle Engineering** The design, manufacture, and operation of motorcycles, automobiles, and trucks and their respective engineering subsystems.

**Vehicle Segment** Classification schemes that are used to regulate, describe, and categorize vehicles.

**Whalebone Buck (Seating Buck)** Sometimes referred to as a packaging buck; an accurate representation of only the vehicle's interior which may include the seats, pedals, instruments, steering wheel, doors, and floor in order to evaluate ergonomics, comfort, and function.

**Zero Emissions** Refers to an engine, motor, process, or other energy source, that emits no waste products that pollute the environment or disrupt the climate.

# BIBLIOGRAPHY

Booker, Christopher. *The Seven Basic Plots: Why We Tell Stories.* London: A&C Black, 2004.

Campbell, Joseph. *The Hero with a Thousand Faces.* San Francisco: New World Library, 2008.

Cheskin, Louis. *Why People Buy: Motivation Research and Its Successful Application.* New York: Ig Publishing, Incorporated, 1959.

Damasio, Antonio. *Descartes' Error: Emotion, Reason and the Human Brain.* New York: Quill, 2000.

Diller, Steve, Shedroff, Nathan, and Rhea, Darrel. *Making Meaning: How Successful Businesses Deliver Meaningful Customer Experiences,* Berkeley, CA: New Riders, 2005.

Gibson, James J. *The Ecological Approach to Visual Perception.* London: Routledge, 1986.

Gibson, William. *Neuromancer.* New York: HarperCollins, 2011.

IDEO. *Design Kit: The Field Guide to Human-Centered Design.* IDEO.org

Jung, Carl. *The Archetypes and the Collective Unconscious (Jung's Collected Works vol. 9a).* Princeton, NJ: Princeton University Press, 1959.

Jung, Carl. *Man and His Symbols.* New York: Dell Publishing, 1968.

Kahneman, Daniel, Slovic, Paul, and Tversky, Amos. *Judgment under Uncertainty: Heuristics and Biases.* Cambridge: Cambridge University Press, 1st edn, 1982.

Lakoff, George and Johnson, Mark. *Metaphors We Live By.* Chicago: University of Chicago Press, 2nd edn, 2003.

LeDoux, Joseph. *The Emotional Brain: The Mysterious Underpinnings of Emotional Life.* New York: Simon & Schuster, 1998.

Lovell, Sophie. *Dieter Rams: As Little Design as Possible.* London: Phaidon Press, 2011.

Macey, Stuart and Wardle, Geoff. *H-Point: The Fundamentals of Car Design and Packaging.* Los Angeles: Art Center College of Design, 2009.

Maslow, Abraham. A theory of human motivation. *Psychological Review,* 50(4) (1943): 370–96.

McKinsey & Company. *Automotive Revolution–Perspective Towards 2030: How the Convergence of Disruptive Technology-Driven Trends Could Transform the Auto Industry.* Advanced Industries, 2016. Available at: www.mckinsey.de/files/automotive_revolution_perspective_towards_2030.pdf

Merriam-Webster. *Dictionary.* Available at: www.merriam-webster.com

Norman, Don. *The Design of Everyday Things.* New York: Basic Books, 2002.

Rainie, Lee. *How will the Internet of Things look by 2025?* (Pew Research Center 2016). Available at: www.pewinternet.org/2016/03/29/how-will-the-internet-of-things-look-by-2025/

*The Rise of the Sharing Economy* (Thinkers 50, 2013). Available at: thinkers50.com/wp-content/uploads/The-Rise-of-the-Sharing-Economy.pdf

United Nations department of economic and social affairs – *World's population increasingly urban with more than half living in urban areas.* Available at: www.un.org/en/development/desa/news/population/world-urbanization-prospects-2014.html

www.daimler.com/company/strategy/ Web (accessed March 2016).

www.mazda.com/en/about/vision/ Web (accessed March 2016).

# INDEX

T - #1016 - 101024 - C240 - 246/189/14 - PB - 9781138685604 - Matt Lamination